MW01200675

THE KINDEST GARDEN

THE KINDEST GARDEN

A PRACTICAL GUIDE TO REGENERATIVE GARDENING

MARIAN BOSWALL

PHOTOGRAPHY BY
JASON INGRAM

INTRODUCTION

> *If you believe in intuitive insight,*
> *the road will open of its own accord.*
> *Masanobu Fukuoka*

When I was four, I camped out one night in a garden den with my big sister. I woke to darkness and silence. Frightened, I peered out of the den. And there saw my grandmother, sleeping in a chair among the roses. A rug over her knees, she was watching over us. Her kind presence transformed the garden. It became an enchanted place full of ancient sounds and mysterious scents. I became lost in wonder, enthralled by the magic of every detail. That feeling of connection, kindness and being totally absorbed in my surroundings is something I'd like to feel in every garden I design.

When I was ten, I read *The Day of the Triffids*, a terrifying science fiction book about giant genetically modified plants that take over the world. The most frightening part was what happened to the survivors. Without any infrastructure left, they had to start a new civilisation. I realised that I didn't even know how to grow food. That began a life-long fascination with growing – and curiosity about our interdependence.

Fast forward many years in the pursuit of achievement and validation, an Oxford degree and a demanding career as an international management consultant, my life imploded when my second daughter died. I sought refuge in the place of deep kinship and connection I'd first experienced with my grandmother. I focused on making a garden. I studied horticulture, garden history and design, then landscape architecture. Over the next twenty years, my children, my garden and my landscape architecture practice grew, together

The pace of learning is picking up and there is a bubbling hope in our collective consciousness.

with articles, talks and my first book, *Sustainable Garden*.

I chose to focus on creating places that were beautiful, but more than skin deep. I wanted my studio to design landscapes for connection and to plant gardens for joy, as well as wider landscapes that linked soil, plants, water, biodiversity havens, corridors and refuges. I gardened biodynamically at home, and our estate masterplans designed holistic networks through farmland, yet I tended not to focus on the way our food crops were grown. Then, after a talk I gave on land energy, a large-scale grower came up to me with tears in her eyes to tell me that my talk had deeply resonated with her. She had her soil tested every year to measure how much fertiliser to add, to raise thousands of hectares of vegetables for supermarkets. 'This year they said my land is basically dead,' she told me. 'We are growing this food in a chemical soup. We buy more and more inputs, it costs a fortune and I can't see where it's going. What do I do?'

I was reminded of *Heidi*, the children's story by Johanna Spyri, where she picks strawberries for her blind grandfather but is persuaded to sell them and takes him money instead. He bites on a coin and cries out in disgust and disappointment. Good food is so much more than a commodity to be chemically optimised for sale.

The grower's question reawakened in me a need to focus on how food should be grown, but at scale. I delved deep into soil, agroecology, mindset and pattern thinking. I studied with Dr Elaine Ingham and Nicole Masters, read piles of books, visited and talked with many generous people at the forefront of newly emerging and ancient best practice. I also focused my studio's work on regenerative landscapes, working with some inspiring, like-minded clients. Our collective issues related to health, adapting to changing weather patterns and feeling the loss of wildlife made this investment an easy decision. My inner work has been as important as my outer work, and both are ongoing. There is also a wider mindset shift happening. It's a great time to be alive and involved, a time when many people are working to put the eco back into economy.

This book comes from the work I have done so far, and a desire to share knowledge between gardeners, designers and growers at all scales. *The Kindest Garden* is about creating beautiful places where whole ecosystems can thrive; it's about moving beyond an extractive system by putting back, nurturing ourselves and regenerating our surroundings. I hope you will take some optimism, some tools and practical ideas from the examples and the approach I'm sharing – together with some inspiration and questions to continue exploring.

There are many imponderables to solve, and the more receptive we become the more questions we raise. What lies beneath the ocean and how do the planets affect us; how best is it to communicate with and coexist with the other beings on Earth and with ourselves; how many millions more unknown fungi, bacteria and viruses are there? These questions all need researchers, and practitioners, with open minds. The pace of learning is picking up and there is a bubbling hope in our collective consciousness.

It's a great time to be alive and involved, a time when many people are working to put the eco back into economy.

WHAT IS REGENERATIVE GARDENING?

> *When you let go of who you are,*
> *you become who you might be.*
> *Rumi*

A regenerative garden isn't just a place, or a method, to grow plants. It's a mindset. It includes and goes beyond sustainability, focusing on making healthier land and people, and particularly healthier soil, which feeds all other ecosystems.

Charles Massy in *The Call of the Reed Warbler* talks about the five landscape functions: the carbon cycle (solar energy); the soil system; the water cycle; biodiversity; and the human community ecosystem. Regenerative growers focus on all of these, and mix in learning from organic, biodynamic and permaculture approaches. It's a movement in a high-energy, start-up phase that is gathering momentum and gravitas as it gains traction. This currently favours approach and intention over rules and certification, sharing resources, experimenting and pooling results according to context, without trying to roll out a one-size-fits-all solution.

Industrial growing

It's easy to forget that industrial farming began with good intentions and pride in scientific discovery. There was a sense that nature could be leveraged without limit and without side effects, to sustain an ever-expanding urban population. Growing food at scale developed gradually from the sixteenth century in Europe and took off after 1909, when scientist Fritz Haber discovered a way to extract nitrogen from the air and fix it in solid soluble form as nitrates – a kind of salt applied to soil as fertiliser. Nitrogen is one of the three key macronutrients that fuel plant growth, along with potassium and phosphate, and is available in soil, organic matter and minerals, respectively. Before Haber's discovery, extra nitrogen was traditionally added to the soil via animal manure (or human 'night soil'), mined minerals and through growing legumes, which fix the nitrogen from the atmosphere in the soil through their roots and symbiotic bacteria.

Nitrogen fertilisers were created to solve hunger and shortages after the two world wars, and they have delivered huge production to feed enormous populations ever since. They have addressed many human problems, and the global payback took time to be understood.

Over time we have discovered that the issues with the use of nitrate fertiliser are many. Nitrous oxide (a greenhouse gas) is produced when it is made, with fertilisers accounting for 6 per cent of greenhouse gas emissions. Nitrate is also a form of salt that kills many of the creatures in the soil that we call the soil microbiome. Once the microbes in the soil are killed, the plants no longer have access to the nutrients the microbes provide, and so like a drug addict they become ever more dependent on their chemical fix of nitrogen alone, and susceptible to diseases caused by the lack of others. Another issue is that nitrates are highly soluble and leech out of soil and into watercourses, where they can kill fish and cause eutrophication – the noxious green algal bloom responsible for blue baby syndrome. High nitrate levels are also a proven carcinogen in food.

While Haber was inventing the nitrogen-fixing process, Paul Hermann Müller discovered that diffusion tensor tractography (DTT), which is derived from nerve gas, killed insects en

As Marina O'Connell, founder of the biodynamic Apricot Centre asked: How might a more sustainable future look?

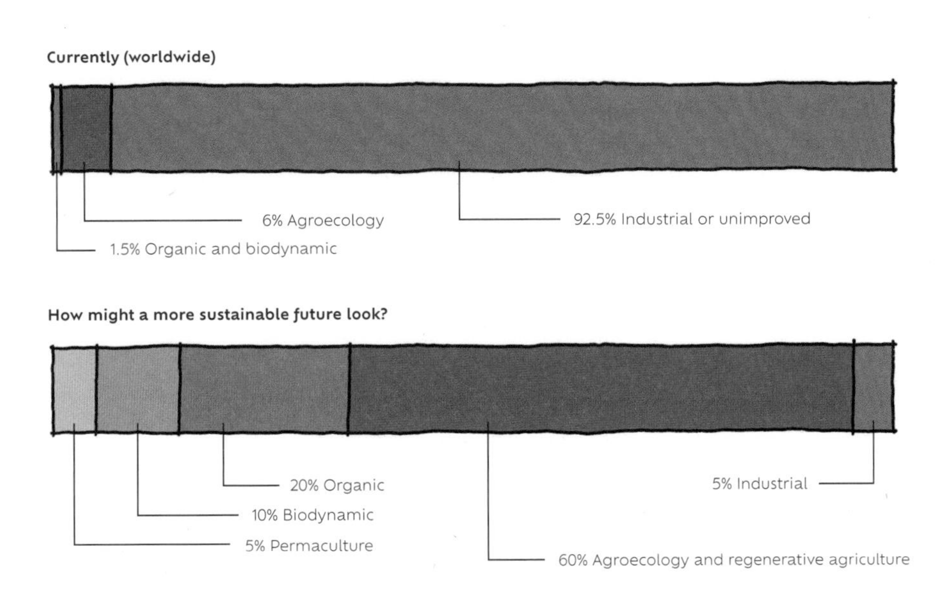

Systems like organic, biodynamic, agroecological and permaculture growing were developed to challenge and resist industrial farming.

masse by interrupting their nervous systems. Together these discoveries enabled food production to become industrial in scale, but they degraded the soil, killing the insects and damaging mammals and the atmosphere in the process. The drive for yield and the introduction of tractors and new machinery also reduced numbers of mixed farms with livestock. Dairy cows were increasingly farmed in sheds, which created issues with their waste becoming mixed. This slurry was spread on the land as fertiliser and ended up polluting watercourses.

Other novelties compounded these problems. To grow plants at scale reliably, seed companies developed hybrid F1 seeds in the 1930s and 1940s. Because they are designed to be sterile and not need pollinators, they don't reproduce naturally so they have created a dependency on seed companies, and disrupted natural pollination cycles. At the same time, paraquat (now banned) and glyphosate were created. Glyphosate kills plants from the roots up. Being a 'systemic' weedkiller, it stays in the plant's system, and has been shown to enter the food system when the plant is eaten by cows, and go into our guts in their milk.

It has taken time for us all to realise how these practices

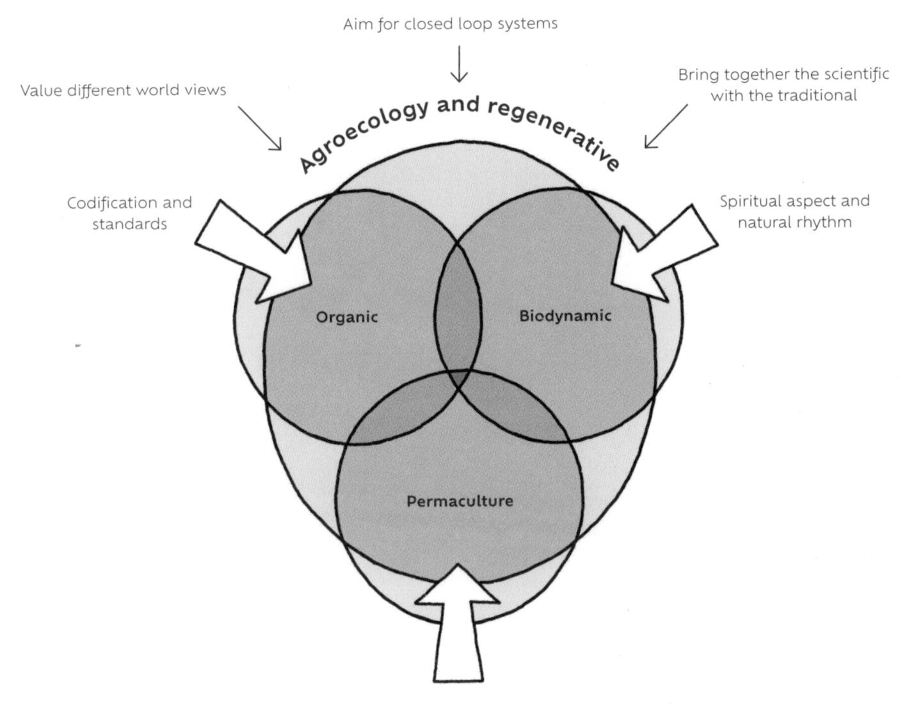

The core principles of different systems, our common ground and how regenerative growers can be inspired by each one.

have been affecting our health and the health of our land. The steady reduction in birds, and the continued loss of insects since Rachel Carson wrote *Silent Spring* in 1962, have recently made it hard to ignore, and no one is quite sure where the shifting baseline was before that.

Hope through regenerative practices

Amid the dominance of industrial agriculture, a parallel movement in smaller-scale agriculture has persisted, with farms below 2 hectares (5 acres) producing 30 per cent of the world's food. Systems like organic, biodynamic, agroecological and permaculture growing were developed to challenge and resist industrial farming, offering well-documented practices and varying levels of training and certification all over the world.

Regenerative growing is the synthesis of these approaches. Practised by gardeners, smallholders and at scale in different regions by pioneers like Allan Savory, John Kempf, Dr Elaine Ingham, Nicole Masters and the Wildfarmed movement, it aims to regenerate entire ecosystems. Its overarching principles include holistic ecosystem improvement, context-specific design, fair and reciprocal stakeholder relationships and continual development of individuals, farms and communities.

Above: While each garden alone may seem small, it can play a pivotal role in this regenerative movement

Opposite: Regenerative gardening offers us a way to bring multidimensional life back into our gardens

Regenerative gardens

While each garden alone may seem small, it can play a pivotal role in this regenerative movement. The fact that there are some twenty-three million gardeners in the UK alone refutes any idea of gardens being irrelevant. Media, advertising and big business have allowed us to divorce growing food from the life cycle of the land, limiting the culture of the garden to a place of exclusively human pleasure that functions in two dimensions only – as an outside decorating opportunity.

Regenerative gardening offers us a way to bring multidimensional life back into our gardens and realise that everything is food to someone, including ourselves. It is an approach that fosters biodiversity from the soil up, includes habitats throughout, increases carbon sequestration in our soils and trees, adapts to climate shifts and develops closed-loop systems. Key practices involve intentional design, mindful material usage, increasing biodiversity from soil to gut, harmonising with natural energies and conscientious plant sourcing.

You don't need a private garden to be part of the shift. Helping at a community garden or allotment can do huge good, as does joining a scheme like Sprout Club – an app linking people offering space to grow with people who want to grow. And those short of time can join a regular box scheme like Riverford, Farm Direct or Abel & Cole, or buy direct from growers and bakers, to eat seasonal healthy produce with fewer land miles and no chemical inputs.

BEST PRACTICES GARDENERS CAN ADOPT FROM REGENERATIVE AGRICULTURE

Regenerative Practice	Practical Applications on a Garden Scale
Holistic Land Management	Rotate vegetable crops, use cover cropping and minimise digging to improve soil health and structure. Include a variety of plant communities to attract beneficial insects and support soil microorganisms. Use mulch to retain moisture and prevent erosion.
Agroforestry	Plant fruit and nut trees alongside vegetables and flowers to create a diverse and productive garden. Incorporate shrubs or hedges as windbreaks and to attract pollinators. Use vertical space by growing climbers.
Rotational Grazing	Mow areas of lawn and wildflower meadows at different times and leave at different heights to mimic grazing patterns. Rotate the location of chicken coops or rabbit hutches to spread manure evenly across the garden for natural fertilisation.
Composting and Soil Regeneration	Use compost bins, wormery and bokashi bins to recycle kitchen scraps and garden waste into nutrient-rich soil amendments. Make biocomplete compost tea and use worm composting to enrich soil and improve plant health.
Water Management	Install water butts or a simple rainwater harvesting system to collect water for garden irrigation. Design swales and raised borders to capture and direct rainwater to where it's needed most. Mulch to reduce water evaporation and maintain soil moisture.
Native Species Restoration	Plant native flowers, shrubs and trees to attract local pollinators and beneficial insects. Create habitats like hedgerows, log piles, bee hotels and sand boxes to encourage biodiversity. Leave areas for detritivores (organisms that consume decaying matter).

TAKING INSPIRATION FROM LARGE-SCALE REGENERATIVE PROJECTS

> *Let the beauty of what you love be what you do.*
> *There are a hundred ways to kneel and kiss the ground.*
> Rumi

Rewilding landscapes, reviving rivers and reintroducing lost species – these projects hold a romantic allure that is hard to resist. The prospect of repairing nature to restore what we've unwittingly erased in our endless pursuit of abundance can be exhilarating and transformative. Headline ventures like Knepp Rewilding and the River Otter Beaver Reintroduction have been well publicised and have stirred strong emotions, from delight to apprehension and outrage. These iconic projects, like a fashion catwalk, showcase what's possible and provide courageous inspiration. Like the best Vivienne Westwood, the idea is not to copy them exactly. We can adopt the best ideas, be inspired by their essence, and create something perhaps more modest in ambition but as transformative in our own context. Alongside the headline projects there are hundreds of farmers and growers embracing organic, regenerative or rewilding systems, and each of these has lessons we can learn from and apply to our gardens.

A beaver dam slows the flow and cleans water while creating habitat for huge numbers of creatures.

Every large-scale project needs a foundational survey of its status quo – a baseline. This involves understanding the land, the soil conditions, what is growing and who lives there – the flora and fauna. Don't wait for baseline measurements to begin, though! Take the example of Knepp, which started inadvertently as the Burrells realised that traditional farming wasn't financially sustainable on their heavy clay. They then left areas to rewild naturally, and the impact they began to witness was awe-inspiring. As gardeners on all scales, we can measure our impact – see page 190 for some ideas of where to begin.

Above: Designing places to restore what we've unwittingly erased can be exhilarating and transformative.

Overleaf: Setting an intention is pivotal, and a good way of encapsulating this is in a master plan.

Setting an intention is pivotal, and a good way of encapsulating this is in a master plan. This involves mapping hotspots where biodiversity thrives, together with corridors and links for creatures to move between them based on existing features – a process that works for any size landscape or garden. The key is to map out the characteristics and uses of each area and different habitats according to factors like wind, sun, soil, landform and water, and see how they can connect and flow synergistically.

When Seth and Lara Tabatznik founded 42 Acres in Somerset, their intention was to create a regenerative retreat centre that would feed both mind and body, following the principles of *Soil, Soul, Society* of friend and mentor Satish Kumar. The name 42 Acres is a light-hearted reference to 'the answer to life, the universe and everything' from the popular 1980s *Hitchhiker's Guide to the Galaxy*. In 75 actual hectares (180 acres) of the magical Selwood Forest was a 2.5-hectare (6-acre) lake. This is now a wetland teeming with life, from swallows to kingfishers and owls, thanks mainly to a family of beavers that have built dams and extended habitat across 3.25 hectares

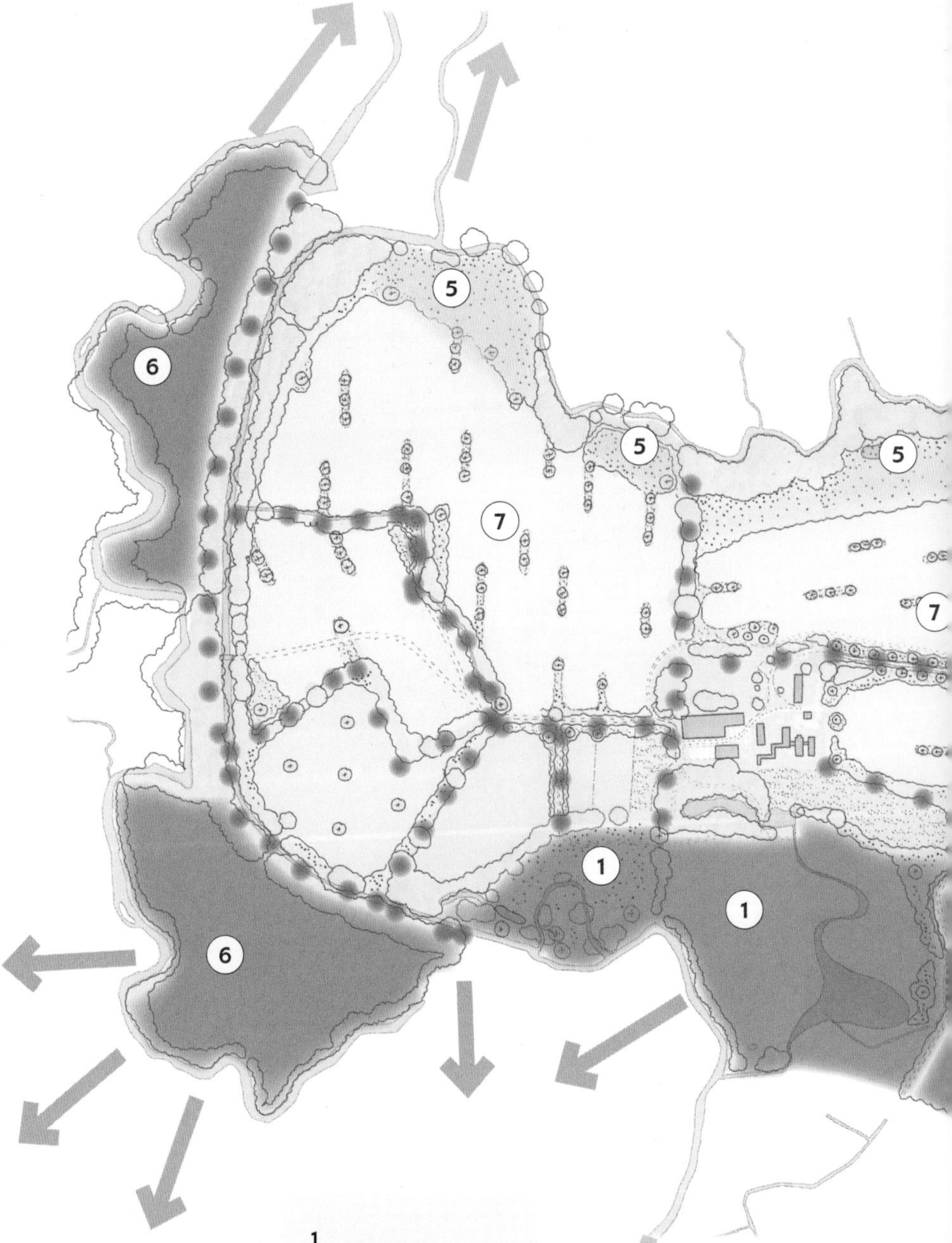

6

FLOODPLAIN MEADOW

- Potential for creation of wet woodland margin and reconnection to old river

7

ARABLE/GRAZING

- Introduce silviculture strips
- Avoid pesticides

1

FLOODPLAIN MEADOWS

- Reconnect to river, restore to include scrapes, ponds and increased water flow
- Potential hay cut or rushes

KEY

Biodiversity hotspot

Biodiversity stepping stone

Biodiversity corridor

Potential connection to adjacent habitat

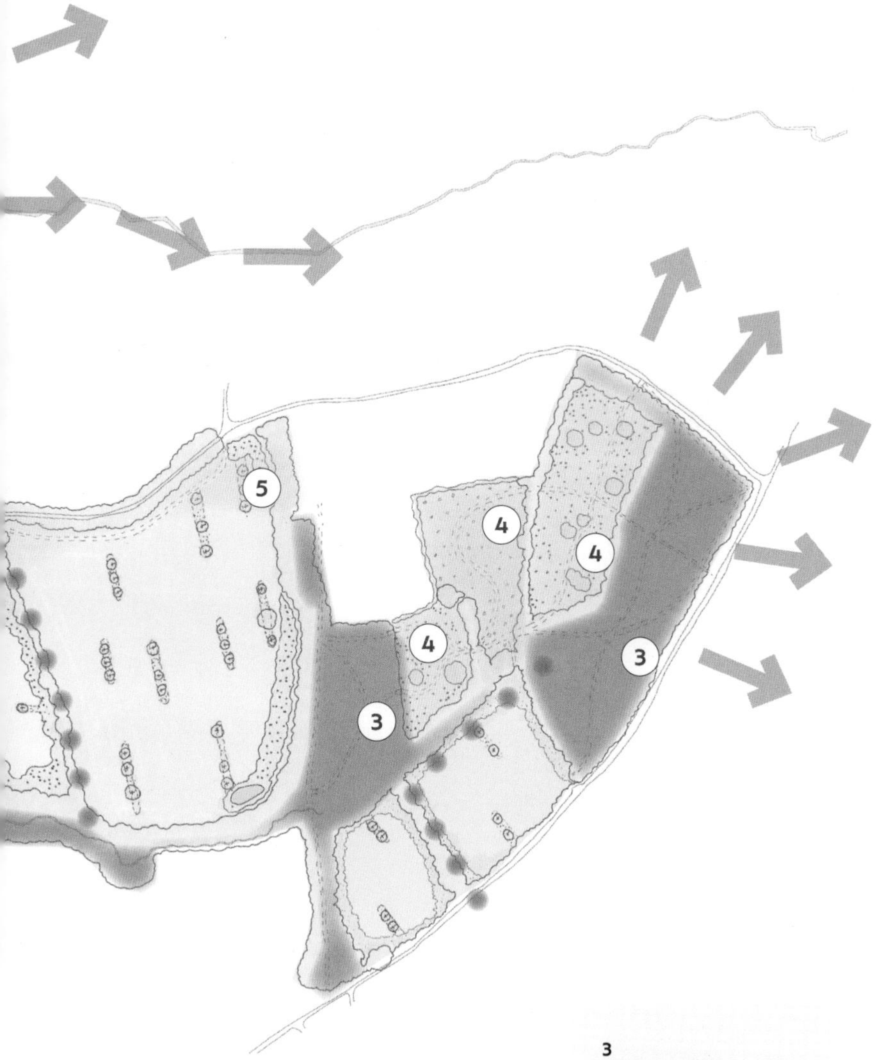

BIODIVERSITY CORRIDORS

- A-frame hedge management and new trees in hedges
- Widen vegetated riverbanks and buffer zone

4
GRASSLAND

- Potential regeneration to woodland

5
FIELD MARGINS

- Increase scrub in margins
- Add ponds
- A-frame hedge management
- Avoid pesticides

3
WOODLAND

- Woodland management plan to increase biodiversity and potentially productive crop e.g. timber & biochar

2
ANCIENT WOODLAND

- Woodland management for biodiversity and conservation

N

Below: The University of Exeter Beaver Wetlands study in Devon.

Opposite: 42 Acres kitchen garden, where nutritious food is grown with minimal food miles.

(8 acres). The retreat centre is fed from a walled garden that grows food curated for the health benefits and tastes the chefs want to add to their menus. Food miles are reduced by growing sweet ingredients like stevia and beetroot instead of using imported sugar cane; exotic lemons are replaced by lemon balm or lemon verbena. Mushrooms are grown on logs and forage plants added to hedgerows and fields. The peace and nurturing intention are evident from the materials used through to the management style. The estate team join retreats and meet every morning to 'check in' with each other. Key team members include a head of 'forage and food' and a land energy expert.

The wider team of stakeholders are a key part of any project – from planning departments to neighbours, environmental agencies and, on a smaller scale, family members. All may have differing opinions and views on what makes a lovely place. In case of overwhelm, keep grounded by referring to your key insights – who is the garden for? What is the kindest thing you can do for your land and yourself?

Fundamental to the regenerative approach is to be allowed to 'fail'. Experimentation and iteration are vital. It was through

Plants like sorrel can provide the citrus taste without the food miles.

'failing' at farming that Knepp was created. It was through abandoning the chemical approach of the 1960s that Riverford was formed, and through iterative trials that the Wildfarmed movement for growing grains with beans and cover crops gained traction. In gardening, a plant death or composting setback also offers a learning opportunity, and each lesson becomes a stepping stone to success.

Sometimes an idea is too good to keep quiet. Marina O'Connell's Apricot Centre grew from a small biodynamic flower and veg box initiative to offering education and therapeutic placements to 2,000 people a year. Similarly, Donna Cox's Moor Meadows in Devon is a testament to collaboration. She manages her meadows through conservation grazing with a micro-herd of native cattle and has created a community to share knowledge, resources and seed. We can learn so much by joining local or online communities. Another perspective on an issue can be a timesaver or sometimes a lifesaver.

The regenerative ethos transcends different scales, but it is not one size fits all. Every piece of land is unique, and understanding local context is a key part of the process. Our current flawed food system stems from disconnecting ourselves from the decision-making process to 'scale up' and do things as 'efficiently' as possible, yet often at the expense of efficacy. To regain stewardship and kindness, spend time getting to know where your food comes from, how it is grown, what goes into the land and into the food – and if you are happy with that. If you have space to grow food yourself, you can get to know your land intimately and join in a process as rewarding as it is ancient. There are a hundred ways to kiss the ground!

Balancing these shifts takes time. The regenerative approach is a steady one and is sometimes referred to as a 'methadone approach' to reducing chemical inputs. It advocates incremental changes. A grower can significantly reduce use of blanket-applied fertiliser by testing the sap in the leaves and responsively balancing deficiencies. A gardener can replace peat-based compost with living compost and learn to make it at home. Listening to the land and to your own energy is paramount, so that you can be kind to both.

Lessons to share:
- establish a baseline;
- plan for the long term;
- approach change with courage and curiosity;
- embrace incremental progress;
- learn through trial and error;
- share knowledge and seek advice;
- embrace setbacks and failure as part of growth;
- look to the long game.

PRACTISING KINDNESS

> *The essence of loving kindness is being able to offer happiness . . . You can't offer happiness until you have it for yourself. So build a home . . . by accepting yourself and learning to love and heal yourself.*
> Thích Nhất Hạnh

What is the kindest thing you could do for yourself?
So much of the current discussion about climate and biodiversity is about how we have undermined the system that serves us. We hear about how this will affect us materially and how uncomfortable and untenable life may become. People discuss ecosystem services, measure what animals, trees and plants can do to support us, and discuss whether it might be more efficient to replace them in part, perhaps with machines or chemicals. A purely scientific utilitarian approach, however, misses the point. This is not just a scientific problem. By trying to separate and control nature we are forgetting our connection, our kinship, with the land and the creatures that share it, and each other.

By embracing both the intuitive and the empirical, we can benefit from ancient wisdom as well as modern understanding.

The word kind and kin both come from the old English *cynd* or *gecynde*, meaning natural, native, of the natural order of things, with a nuance of taking care of one's own. Being kind is naturally helpful to our survival. Understanding things through a scientific lens is helpful too. Yet the less reductionist truths, which we all know at some level, are not only scientific. They are also based on a more profound, heuristic knowledge – a blend of *mythos* (narrative) and *logos* (reason). By embracing both the intuitive and the empirical, we can benefit from ancient wisdom as well as modern understanding.

Humans have grappled with the 'why' for millennia. All major philosophies explore routes to understanding our connection to something bigger than ourselves, whether they call it the Universe, Gaia or God. Early Chinese philosophers spoke of qi – which is not a god or a being but the energy or life force that pervades everything and links the plant, animal, human and divine world. Christianity teaches the sacred as being beyond human, and Confucian scholars like Tu Weiming see humans as co-creators of the universe: as anthropocosmic, neither privileged nor unique, but together with the *wanwu* (the myriad other beings in nature) forming one body. The Golden Rule – to do unto others as we would be done by – applies to all. Similarly, in the sixth century BCE, philosopher Laozi defined the Dao. He described it as a process, the 'Way' or unknowable force that keeps the world in being and animates everything that exists. The creative energy of the Dao is cyclical, with all creatures returning to their roots in stillness (*jing*) rather than a final death, before springing to life again.

These philosophers suggested that we observe the Dao in nature, and immerse ourselves in the natural world to overcome our destructive egos. Both Dao and qi are similar to what Christian philosopher Thomas Aquinas called 'Being itself'. An omnipresent power or *Rta*, 'the way things truly are', was also described by the ancient Vedas – sacred Hindu texts and powerful poetry. This is furthered in the concept of the 'Buddha-Nature', described by Zen master Dōgen Kigen, as the same sacred potential that all things human and non-human already are, with nothing to do but realise it. Sufi teachings mirror the same reverence: humans are not placed above nature but are intrinsic to it – a teaching that is integral but sometimes overlooked in the modern Christian tradition.

All these approaches through thousands of years recognise that humans have an inherent kinship with all beings on Earth. Even in our most secular circles we see this, in our perennial fascination with wildlife programmes and our affection for baby animals; this is something street artist Banksy recently played on when he painted a giant kitten on a wall in a war-torn setting, suggesting that would draw more attention than the conflict alone. How then can we 'live according to nature' like the cosmopolitan Stoic philosopher Hierocles of the second century CE? We can begin with nurturing ourselves and extend outwards, embracing all life forms as we recognise that we are all kin; we are all connected.

Kindness to ourselves and to others is rooted in empathy, the ability to understand and share feelings. Caring for a garden begins with trying to understand its language. We usually start by getting to know the trees and plants. It's a joy to begin to notice which plants like to grow where and with which other plants. It is also a revelation to learn which birds visit the garden,

Above: Humans have an inherent kinship with all beings on Earth.

Opposite: Understanding a garden means viewing it through concentric circles of kindness or concern starting with yourself.

what songs they sing and what they eat. If you haven't already, watch where bees and butterflies land, where they feed, drink and sleep. (Check inside foxgloves for sleeping bumblebees.) Identify earthworms, beetles and woodlice; watch parasitic wasps catching spiders, spiders catching flies and water boatmen skidding across the top of water like ice skaters. Observe your weeds to reveal insights about your needs, telling you about soil compaction or mineral imbalances in your soil and your gut. Learn what is in your food; discover what your food ate. All of this will connect you more deeply to your ecosystem's intricate web and its role within your body and your own health and happiness, and from there kindness can only spread.

So what is the kindest garden? An embrace of beauty, a sanctuary, a place for knowledge, abundance and community – and a mindset most of all. This gentle art and science of landscape gardening fosters kindness towards ourselves, our spiritual and physical environment, and all the beings we share this world with. With this intentional approach the simple act of tending to the land becomes a profound expression of kinship. What if your garden was the kindest thing you could do for yourself?

TOP TEN INSIGHTS FOR REGENERATIVE GARDENERS

> *Like each branch is an integral part of a tree, every creature, human or non-human, is an integral organ of the Earth.*
> *Satish Kumar*

When you think of a garden, what do you think of? A place to relax, play or entertain? A place of reflection and refuge? Of abundance and celebration, or of learning and community? A place to grow flowers or food, or to watch wildlife? Or do you think of it as a burden, a list of chores, of overwhelm or disappointment? A ceaseless battle for order or harmony? It's interesting to notice, without judgement, and to wonder whether how you feel about your garden at different times might reflect how you feel about life in general or yourself.

Here are ten ways to be kind to yourself and your garden. Add your own as you think of them.

1. Who are you gardening for?

- Think of all the beings, human and non-human, that you are gardening for – make sure you are on the list.
- Decide what your garden space will be, for them and for you – how will you know if it is working?

2. Ways to be kind to yourself

- What would you do for, or say to, yourself if you were your own kindest friend, wise grandmother, loving granddaughter or son?
- Can you see your garden as an extension of yourself, a powerful little piece of the world, beyond the notion of boundaries – and extend that kindness out?

3. Ways to be kind to your soil microbes

- Feed them with organic matter and living compost inoculant.
- Use rainwater for watering the garden.
- Nurture them with layers of planting and living roots.
- Protect them from compaction with mulch.
- Disturb them as little as possible so they can thrive.

4. Ways to be kind to your gut microbes

- Eat many different fruit and vegetables; choose colour and scent for secondary metabolites; forage when possible and safe, to add more variety and plant phytonutrients.
- Filter mains water to remove chemicals before drinking.
- Make healthy compost to feed to the soil and your gut; close the loop.
- Avoid processed food and excesses of poisons like

glyphosate, neonicotinoids, chemical fertilisers, emulsifiers or bleaching agents (banned in the UK) – disturb your gut microbes as little as possible so they can thrive.

5. Ways to be kind to your water

- Collect rainwater to use in the garden.
- Aim to limit use to 110 litres (24 gallons) of mains water per person per day.
- Install a compost loo if you have the space.
- Use bio-detergents to clean your house, clothes and body.
- Use chloramine-free rainwater, or add humic acid to tap-water before spreading it in the garden.

6. Ways to extend kindness to trees and all plant beings

- Use open-pollinated seed.
- Avoid genetically modified organism (GMO), neonicotinoid or other 'cide-doused seeds (*cide* is from the Latin for 'to kill').
- Plant in layers and be mindful of the archetypal communities your plants belong to – where might they rather be?
- Think of the whole tree or plant, from tip to roots and the soil around it.
- Protect tree roots and allow them to grow up with 'friends and family': that is, a mycorrhizal network.

7. Ways to be kind to the animals that share your space

- Share your food – avoid '-cides' and grow enough to share with the creatures that coexist.
- Plant alternatives like nasturtiums (*Tropaeolum*) to offer to caterpillars instead of cabbages.
- Use physical barriers like nets for prize plants you don't want to share. Uncover them later in summer to allow pollination when enough predators are around.
- Share your home: allow bees in lime mortar between bricks; allow wood to decay gradually to home insects like leafcutter bees; allow birds to nest under eaves; and give bats and insects crevices or boxes.

8. Ways to take care of your energy

- Solar energy
  - Fossil fuels – use intentionally. Know how much you are consuming and where you might reduce.
  - Energy stored in plants – try to return 70 per cent to the land as biomass and harvest 30 per cent.
  - Energy in the soil: protect the fungal hyphae below ground that amass carbon; minimise digging; and avoid poisoning the soil.
- Wind energy
  - Direct the flow of wind in the garden to help shift a frost or cool a hot summer's eve.
  - Position filters like hedges and trees to deflect wind from channelling into wind tunnels or tree flatteners.
- Land energy
  - Design for flow, for prospect and refuge, for yin and yang, to create healthy spaces for humans and animals.
- Human energy
  - Ground your body and rest your mind regularly.
  - Take care of your gut biome and your outer biome.
  - Keep healthy boundaries and seek connection to the natural world.
  - Find friends that love what you love!

9. Share your knowledge and ask for help

- Join a garden group or forum and organise a seed swap or perennial plant split, or go on visits to learn from other growers.
- Join a regenerative growers' cluster group.
- Share what's in season, what's good to forage and where and when.
- Follow up the links and resources at the end of this book (see pages 230–4), and discuss the ideas.

10. Be the kind of ancestor you'd like to be.

> *The Earth does not belong to us. We belong to the Earth.*
> *Celtic saying*

Use open pollinated seeds, plant in layers and create communities of plants above and below ground.

HOW TO USE THIS BOOK

> *Out beyond ideas of wrongdoing and rightdoing, there is a field. I will meet you there.*
> Rumi

I have been asked to write a book that is accessible, but there is no dumbing down, as this is important work. There is a little gentle science, but gardeners are used to Latin, and *scire* is just Latin for 'to know'.

Please take what you can use and be curious about the rest. For a quick page turn, read the introductions to each part and bookmark any tables and projects that you might want to come back to. I have assumed a certain level of knowledge following on from my previous book, *Sustainable Garden*, and with so much more to know and limited space in one book, there are lots of notes and resources at the end to explore areas that interest you.

Each section in Part One has a diagram of a typical garden with a small front garden and larger back garden to demonstrate the ideas in that part of the book. The size is based on the average UK town garden, and it is meant as a simple visual tool to help readers apply the ideas in the book at different scales. You might take the layouts and either measure your own garden on Google Earth or pace it out in rough metres (yards) to scale up or down in quantities, but choose your own design layout to suit your context, aesthetics, sun, wind, views and energies.

I have included case studies to show that best practice can be relevant to any scale and space. There are tables of information to refer to when planning a garden, and a few simple projects you can follow to experiment with.

Initially, I focus on relationships between us and key elements in the garden. I start with soil, then water, ecosystems, materials, energy and planting. The second part looks at how all of us can work to heal our relationships in practical terms; what we can measure to know if we are doing good or harm. The last part explores how creating a regenerative garden might just be the kindest thing we can do for ourselves. Finally, there is a large appendix with tables and references for further reading. I hope you use this as a springboard for many a deep and fascinating rabbit hole of ideas.

An average sized UK garden with regenerative principles applied.

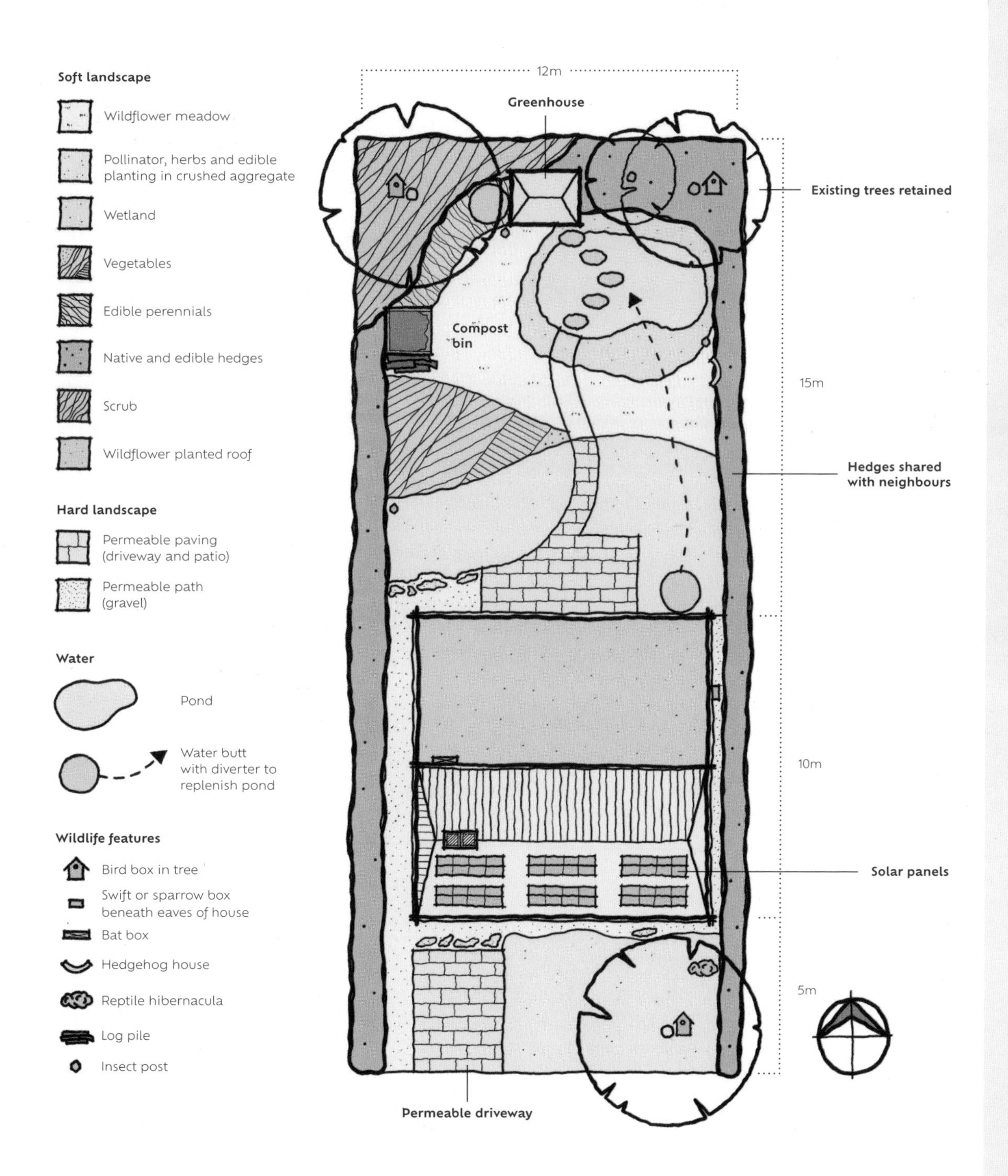
Soft landscape
Wildflower meadow
Pollinator, herbs and edible planting in crushed aggregate
Wetland
Vegetables
Edible perennials
Native and edible hedges
Scrub
Wildflower planted roof
Hard landscape
Permeable paving (driveway and patio)
Permeable path (gravel)
Water
Pond
Water butt with diverter to replenish pond
Wildlife features
Bird box in tree
Swift or sparrow box beneath eaves of house
Bat box
Hedgehog house
Reptile hibernacula
Log pile
Insect post
12m
Greenhouse
Existing trees retained
Compost bin
15m
Hedges shared with neighbours
10m
Solar panels
5m
Permeable driveway

PART ONE

RELATIONSHIPS

THE ELEMENTS OF THE GARDEN

SOIL: WHAT IS IT?

In the garbage, I see a rose.
In the rose, I see compost.
Everything is in transformation.
Impermanence is life.
Thích Nhất Hạnh

Soil contains the alchemy that allows everything on Earth to live, die and create more life in turn. It reminds us that we are both infinite and fleeting. It's our first and last community, our source of life and our place of death. We grow our food in soil and bury our dead there, hide precious treasure in times of danger and secrets in times of shame. We build our houses from bricks, utensils and art from clay, antibiotics from its microbes. As humans we rely on a healthy soil biome to grow nutrient-dense food and keep our own guts healthy; so vital is this to our survival that even the most hardened city dwellers are still naturally hardwired to sniff out healthy soil. Our noses can detect geosmin, which is Greek for 'earth smell' – a compound produced by microbes after rain – at five parts per trillion. This is 200,000 times more sensitive than sharks' ability to smell blood.

Wet earth smell gives us a dose of the hormone serotonin, drawing us to it unconsciously because it makes us feel good. And yet many of us are soil blind, unable to see or appreciate the riotous maelstrom of life under our wheels and feet. Appreciating soil gives us a gateway to understanding how to be kind to our bodies and our Earth. No one has mapped all billions of components of soil yet, but we are collectively beginning to dive deeper into this magical realm. We know soil has parents; it has a lifespan and a life cycle. It starts in the explosions of volcanoes and the grinding of glaciers and can last for millions of years, or be killed in an instant by chemicals. And become dirt.

A garden may have similar base geology and soil texture, but areas will have varied health and microbe levels depending on what has been growing and how soil has been managed.

Soil's parents are rocks, which are broken down by microorganisms, weather and plants to become sand, silt or clay, and nutrient-holding living soil. The nutrients feed roots to grow plants which then feed animals, which in turn die and break down to become soil, or fossils and rocks again.

Our many myths show our deep reliance on soil. In Egyptian mythology Khnum creates human children from clay. In Greek mythology Prometheus moulds men out of water and earth. The Sumerian god Enki makes humankind to serve the gods from clay and blood. In Islam God creates man from clay, and in Christianity God forms man from the dust of the ground and puts Adam (*Adamah* means 'earth') into a garden to tend it.

Our ancestors revered the catalytic properties of soil. The famous herbalist and Benedictine abbess Hildegard of Bingen includes a remedy in her twelfth-century *Physica* for numbness, which involves applying earth to the skin at key points, and another less pleasant remedy for scrofula, which involves eating earthworms' innards to get the benefit of the digested soil. While I don't recommend eating them, wormcasts are the perfect medium for growing plants.

TAKING INSPIRATION FROM THE SOIL NETWORK

> *The soil is the great connector of lives, the source and destination of all . . . without proper care for it we can have no life.*
> *Wendell Berry*

Soil is everything we have been and everything we will become. It transforms what has died into new life. How does it work this magic?

Your garden soil will likely be 90–97 per cent mineral and 3–10 per cent organic matter. The composition of the mineral layer is important because this holds nutrients for your plants. It also determines the texture of your soil and that determines several things, including how much your back will ache if you don't opt for a no-dig garden!

You can search on geological websites like bgs.ac.uk in the UK or usgs.gov in the US to find your local geology, which will tell you whether the underlying rock is igneous or sedimentary, and whether this was changed by temperature or the pressure of a glacier passing to become a harder morphic rock. Over many thousands of years minerals in these rocks are broken down and migrate to the top layer to become part of your 'top' soil together with organic matter. Depending on the rock below and events over millennia, your soil will be predominantly sandy, clay, silty or loamy, plus organic matter.

If you have clay soil you will be able to roll it into a sausage that holds its shape or even make a cup out of it. Because clay is many tiny particles, it has many surfaces and a very high total surface area. This gives clay a high cation-exchange capacity (CEC), meaning it is good at holding on to nutrients and water, and good at exchanging energy. This makes excellent growing soil, as long as you also have a healthy micro-herd of microbes in organic matter to help release the nutrients in the mineral layers. With so many tiny particles, clay soil can become anaerobic (without oxygen) if your good microbes have been killed off through compaction or poison. You can tell clay is anaerobic if it is grey to yellow and smells of sulphur (rotten eggs). If this is the case it will need some help to allow healthy plants to grow.

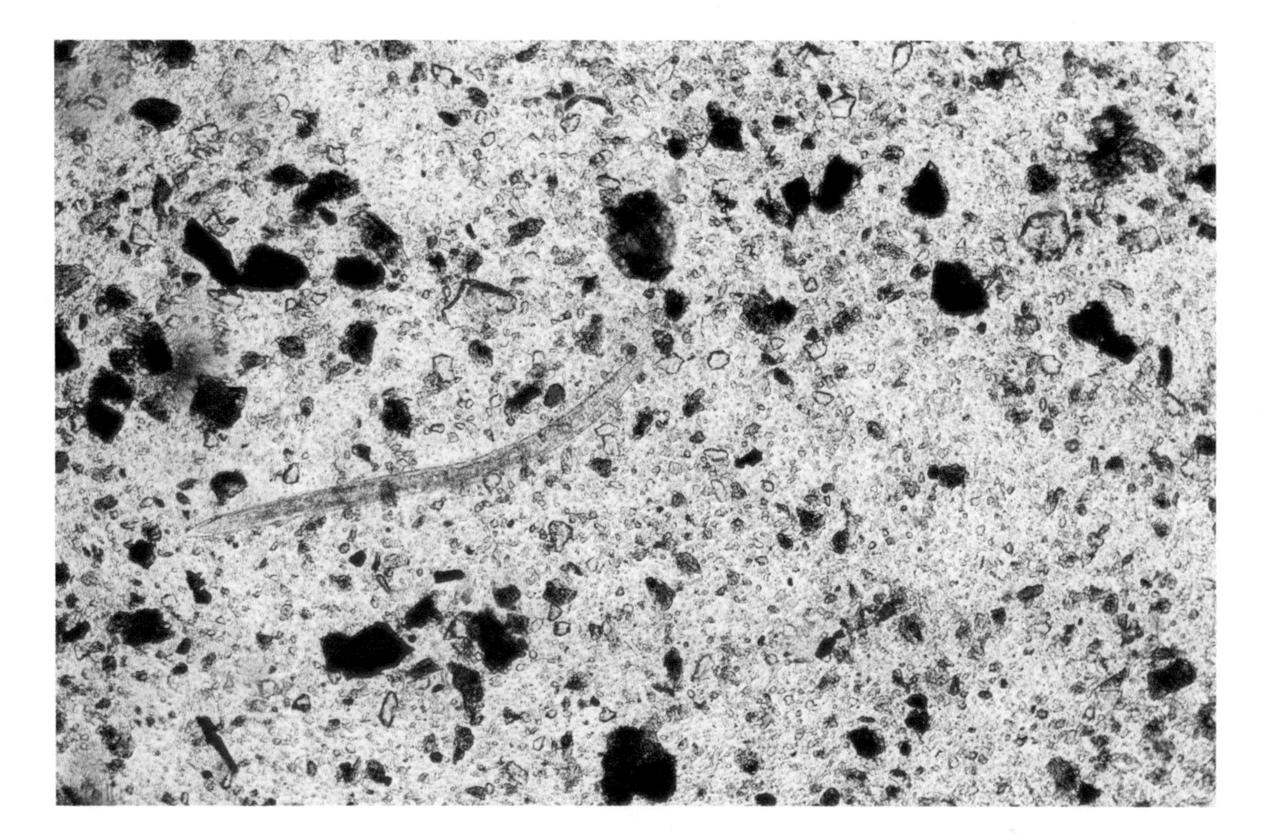

Opposite: The organic matter is where the underground gardeners live.

Above: A nematode is made visible under the microscope.

Sand is the coarsest soil, made up of irregular-shaped particles with big gaps between them. Sand feels gritty when rubbed in your hand, and particles are large enough to see and pick up with tweezers. The spaces allow water and air to flow through quickly, so sandy soil warms up earlier than clay in spring. This allows early seed germination but emits more heat into the atmosphere in hot spells. If there are no organic microbes holding these large particles together, nutrients can wash away and plants dry out quickly.

Other types of soil include silt, washed to a silky texture from riverbeds, and loam, which is a happy mixture of many soils after good organic matter has been added over many years. And this is the key point to take away – for none of these types of soil is good or bad or more or less fertile in itself. Each has fertility in its base minerals, and needs organic matter to unlock the nutrients. Why? Because the organic matter is where the underground gardeners live.

WHO ARE THE UNDERGROUND ALCHEMISTS?

Let's go in size order. If you dig a spit (spadeful) of soil, you should turn up between nine and sixteen earthworms, more in wet and less in very dry, hot or cold weather. We have thirty-one types of worm in the UK. Our largest, the common earthworm, can live four to eight years and grow to 35cm (14in). Its deep vertical burrows provide aeration and take down carbon. The smaller topsoil dwellers like the green worm are a paler greenish colour and burrow horizontally, gluing soil aggregates together with mucilage. Surface dwellers like the little tree worm don't burrow but also start the process of breaking down organic matter into plant-available nutrients.

Worms are the biggest of the underground gardeners that feed on all the smaller microbes in the soil, squeezing out the goodness of their innards to leave a trail of plant-nutritious wormcast. This goodness comes from nematodes and protozoa, fungi, bacteria and viruses that they feed on in turn. The fungi also consume the bacteria, and occasionally the bacteria consume the fungi. It's a web economy! The magic for growers is

The way these microbes work with the plant roots creates food for plants in the waste-free world that is the soil food web.

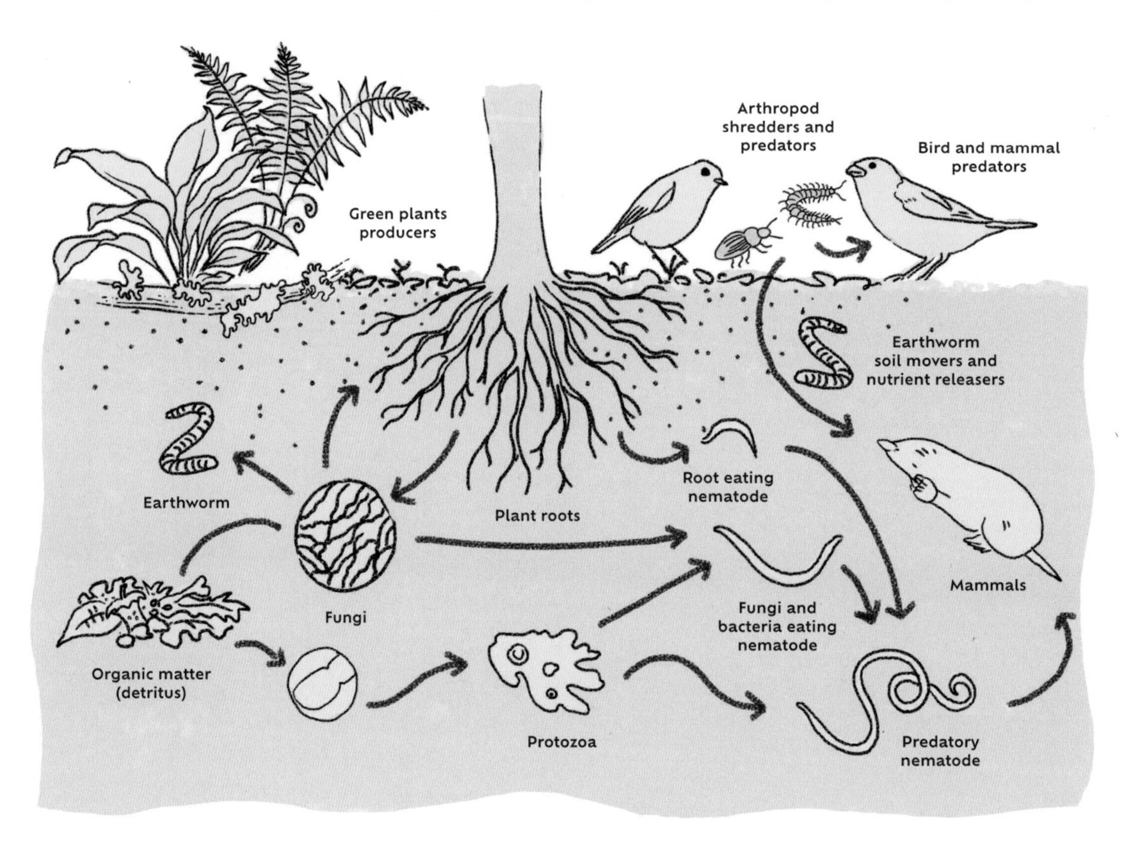

in the way these microbes work with the plant roots to create food for plants in the waste-free world that is soil. The simplified diagram opposite explains how the cycle works. It's worth focusing on the diagram for a moment, as understanding soil makes everything about regenerative growing make sense, and everything about poisoning and killing our soil seem like insanity.

The tree I'm sitting under to write this grows roughly two-thirds above ground and one-third below, and uses the energy of the sun on the chlorophyll of its leaves to store carbon, through the miracle that keeps us all in food – photosynthesis. Incredibly it then sends about 40 per cent of this hard-earned food to its roots and, rather than storing it there, sends some, as exudates, to feed the microbes (fungi, bacteria, protozoa, etc.) that live in the soil around the roots. The food is exchanged in return for a requested specific nutrient each time, such as calcium, magnesium, sulphur or water, which the bacteria 'fetch' from the minerals available in the soil from base rock, and deliver to the root area for the plant to take up. If this sounds like a pizza delivery service the analogy gets better, because legumes (plants in the pea family) often store nitrogen inside aggregates (tiny clumps of soil), which they build around strands of fungi that attach themselves to the roots: a stockroom in human terms! This symbiotic relationship benefits the microbes, which get the exudates, and the plants, which get the water and nutrients, and most of all it benefits us. The healthier the microbes in the soil, the better able they are to provide the plant with the exact nutrients it needs, and the more nutrient-dense the plant becomes, the more health it can give us.

Our understanding of the relationship between bacteria, fungi and plant roots is deepening all the time, with schools like Elaine Ingham's Soil Food Web leading the way. Recent discoveries include plants that exchange nutrients and chemicals not only with the soil microbes but also between each other, such as nitrogen-fixing trees, which share their nitrogen with nearby plants in exchange for minerals and food. This underground web is used to communicate other information too, like the danger of an insect attack, which causes leaves in vulnerable plants to turn bitter, or the forest fire alarm, which tells trees to drop their leaves to help slow the spread.

UNDERSTANDING SOIL IN OUR GARDENS

> *A soft symphony of sounds emanates from the soil within a forest and the more thriving the ecosystem, the greater the diversity of noise.*
> *New Scientist*

How does our soil biology impact our gardens on a practical level?

To answer this, let's look below ground. To have a fully functioning soil microbiome we need enough bacteria, fungi, protozoa and nematodes to feed plants, in the proportion for the type of plants we want to grow, and to minimise the microbes that will harm our plants.

If you want to know how well your soil supports your desired plants you can learn to look at it under a microscope or have it tested by one of an increasing number of dedicated soil technicians.

A soil that is in an early succession phase through being disturbed or damaged is highly bacterial. Bacteria are the early clean-up crew, eating simply and multiplying rapidly. If the soil microbes are more in balance there will also be plenty of protozoa, amoebae and nematodes further up the food chain, and lots of fungi. A typical vegetable garden needs as much fungi in the soil as bacteria, at least 135 micrograms of each per gram of soil. As the plants above ground become more complex and woody, these plants will thrive in more fungal-dominated conditions, wanting a balance of between 1:100 and 1:500 bacteria:fungi for shrubs and a whopping 1:50,000 for old growth forest.

Only early succession plant life will be supported above ground if the micro-herd is predominantly bacteria. If you have frequently disturbed, compacted or poisoned ground or a garden with minimal organic matter, it will mostly support 'weeds', wild flowers and some pioneer shrubs. In a brownfield garden with minimal topsoil the plants grow 'hard' and will be tough, often very floriferous and still excellent for pollinators but not for mammals that need higher nutrient food plants. These gardens have early succession soils as in the diagram on pages 48–9.

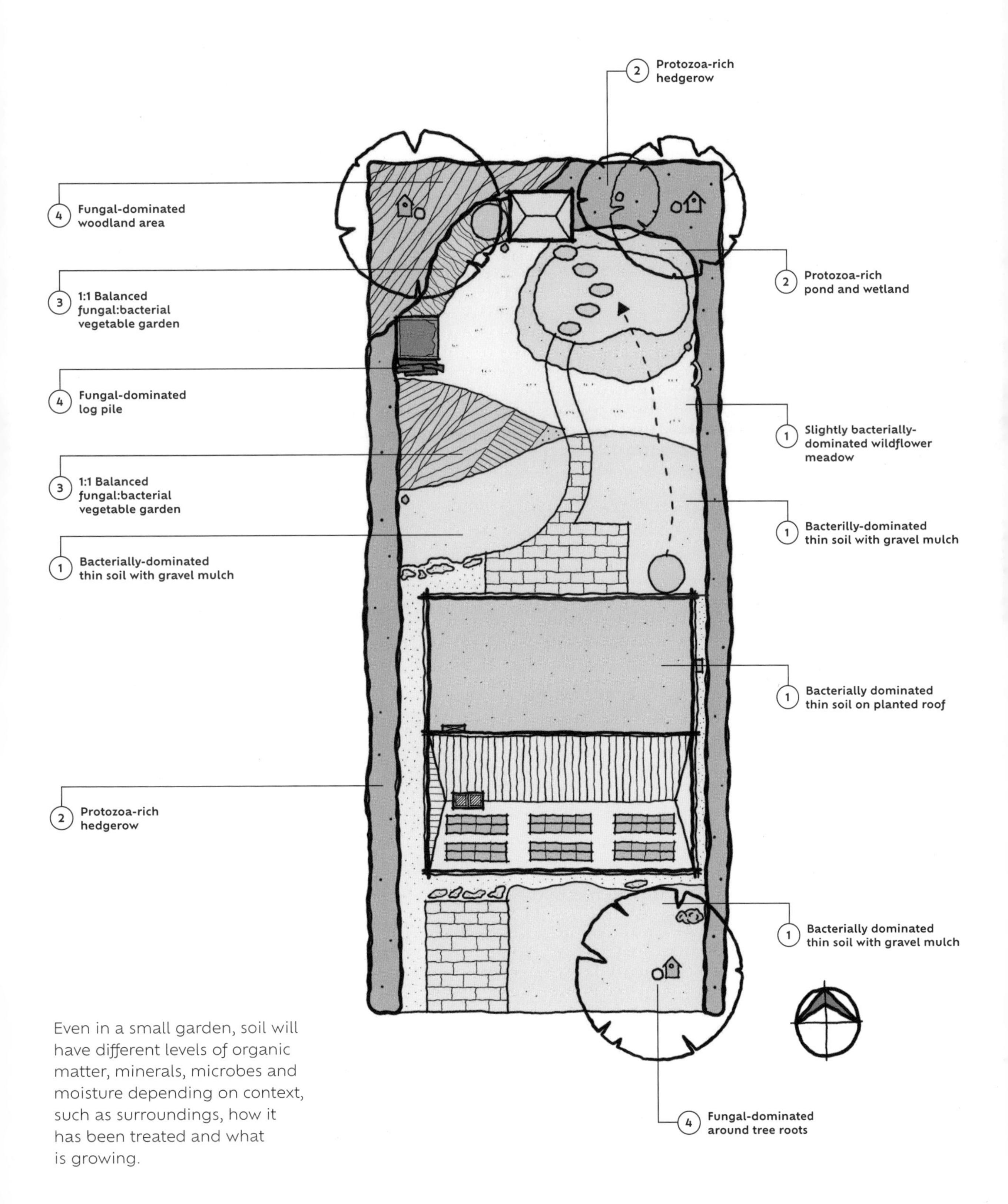

Even in a small garden, soil will have different levels of organic matter, minerals, microbes and moisture depending on context, such as surroundings, how it has been treated and what is growing.

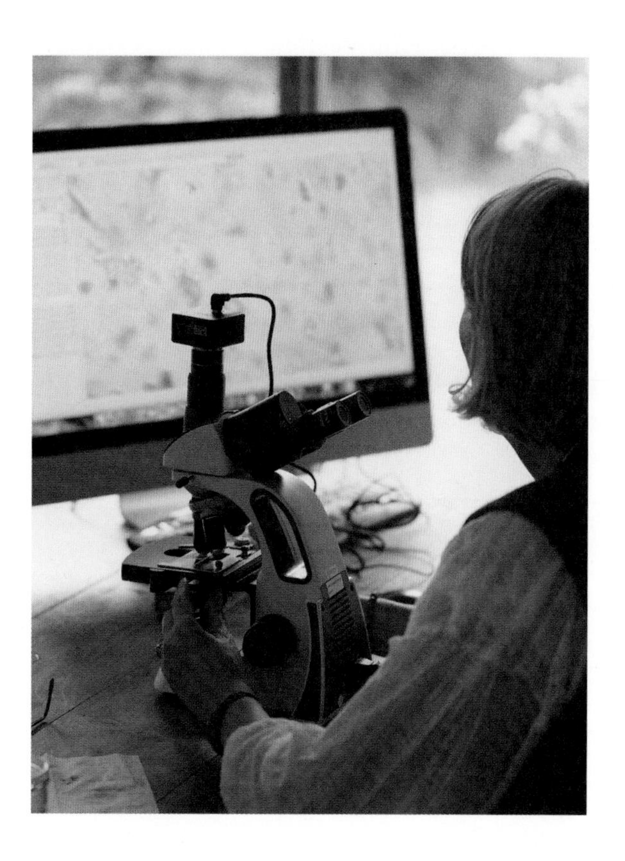

Once you see the crazy, little, Pac-Man-type ciliates or the slow moseying gait of the moss piglet, a cute creature so indestructible that it can survive outer space, you may be hooked!

Many agricultural soils that have been over-tilled, weedkilled and over-fertilised are also highly bacterial and low in other available nutrients. If they have too much manure, they may be very rich but rich in bacteria – what soil guru Nicole Masters calls 'constipated soil'. The nutrients can't get from the soil to the plant without the protozoa, nematodes and microarthropods, so farmers rely on artificial inputs like chemical nitrogen (N), phosphorus (P) and potassium (K) to feed the crop. How much cheaper it would be to create the conditions for healthy roots and allow the plant and a fully diverse herd of microbes to gather what they need.

Some gardeners also rely on chemical fertilisers. These give a quick flush of 'fast food' with N, P and/or K supplied to plants in a quickly available form. Plants may grow fast and soft but the problem is that these fertilisers are made predominantly of salts, which kill the microbes in the soil whose job it is to feed the plants exactly what they need, from a much wider spectrum of minerals and metabolites than N, P and K. This creates a vicious circle, as without the microbes the plants become dependent on fertilisers, making fertiliser sellers richer and our soil and food poorer.

As gardeners we are generally trained to think of pH as the baseline for our soil, to choose plants that thrive in more acid or alkaline soils, or manipulate the pH with lime (to make it more alkaline) or sulphur (more acidic). Once we understand the succession chart overleaf, it makes sense that grassland soils are naturally more alkaline and forests more acidic. When we want to create grass, meadow, vegetable garden or woodland-edge habitats, we can help our soil microbiome supply what our ecosystems need naturally.

How does this work? Plants put out foods, or exudates, through their roots to feed the microbes in return for nutrients. Bacteria consume these exudates and convert them into nitrates, and fungi consume them (humic acids, to be specific) and synthesise ammonium. Nitrates make soils more alkaline, and ammonium makes soils more acidic.

At the beginning of the succession (on the left of the diagram), bacteria count is high and fungi count is low. This means that the most prevalent nutrient is nitrate (NO_3); this is perfect for grasses and 'weeds'. As we move towards more complex systems, there is proportionally higher fungi to bacteria count in soils. This means that the most prevalent nutrient is ammonium (NH_4); therefore the soil is more acidic. What grows best in these soils? Perennials, shrubs and, finally, trees.

Once we understand this, rather than dumping lime or sulphur on to our soil, we can make sure to curate our microbes, feeding our bacterial herd if we need to disturb the soil in earlier succession planting, and increasing our fungal herd if we want later succession plants.

If you want to know how well your soil supports your desired plants, you can learn to look at it under a microscope or have it tested by one of an increasing number of dedicated soil technicians, who will send you a report and describe who does what. They may even share images of your underground herd with you via video link, but beware! Once you see the crazy, little, Pac-Man-type ciliates (a type of protozoa that whizzes around eating all the bacteria it can bump into) or the slow moseying gait of the tardigrade or moss piglet (a creature so cute but indestructible that it can survive outer space) you may be hooked! Soil will never look the same again. To learn more on testing soil see page 50.

SOIL SUCCESSION

To have a fully functioning soil microbiome we need enough bacteria, fungi, protozoa and nematodes in the correct proportion to feed the plants we want to grow. Soils naturally go through a succession of typologies over time from highly bacterial to highly fungal. By understanding where our soils are on this path, we can understand how to work with rather than against nature in our gardens and landscapes.

Succession 1: Bacteria
After a major disturbance like a volcanic eruption, earthquake or fire, rocks are bare and the mineral or parent material is exposed. Bacteria alone are present.

Succession 2: First microbial activity
The first colonisers are bacteria, which can multiply every 20 minutes with sufficient food. These are followed by their predators: protozoa, nematodes, fungi and microarthropods. If there is no repeated disturbance, lichens and mosses will grow and hold water like sponges, begin to decompose and create organic matter.

Succession 3: Weeds
Bacteria-tolerant, early pioneer plants, which are fast growing, such as the seed-dispersing annuals scarlet pimpernel (*Anagallis arvensis*) and chickweed (*Stellaria media*), begin to scramble along the surface. Deep-rooted perennials like bindweed (*Convolvulus*), docks (*Rumex*) and nettles (*Urtica*) send roots down to pull up nutrients. As they decompose, they create nutrients and organic matter for more herbaceous plants.

	BACTERIAL DOMINATED			
FUNGAL:BACTERIAL RATIO	**100% Bacteria**	0.01:1	0.1:1	0.3:1
	Alkaline			
ORDER OF SUCCESSION	**Bare Parent Material**	Cyanobacteria, true bacteria, microarthropods, protozoa, fungi, nematodes	Weeds, high nitrogen	Early grass, brassicas, mustard
		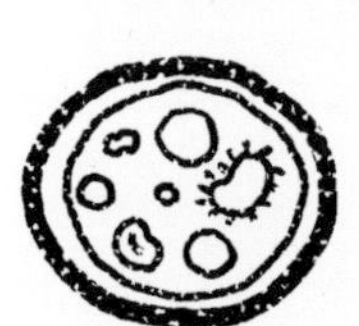		

Succession 4: Early brassicas, mustards and grasses
These are nutrient scavengers that grow fast: they may have deep roots to break up compacted soil. As they die, they make the nitrogen they have accumulated available to other plants like grasses and forbs, with a network of roots beginning to form below ground and a community of insects above ground.

Succession 5: Grasses and perennials
When the fungus to bacteria ratio is 0.75:1 herbaceous plants will dominate and a thick sward will develop.

Succession 6
With equal numbers of fungi and bacteria in the soil, there is an ideal environment for most crops, and such soil is highly productive for horticulture.

Succession 7
As fungi begin to prevail over bacteria in the soil, climbers and shrubs are able to grow and soil becomes gradually more acidic.

Succession 8
Deciduous trees and woodlands begin to thrive once there is a fungi to bacteria ratio between 5:1 and 500:1. Soil is more acidic, and underground fungal networks become established.

Succession 9
Old growth forest and conifers live at the extreme of fungal-dominated soils. These will survive for centuries unless a major disturbance event like fire or landslide returns them to an earlier stage in succession and allows earlier succession species to dominate in some areas.

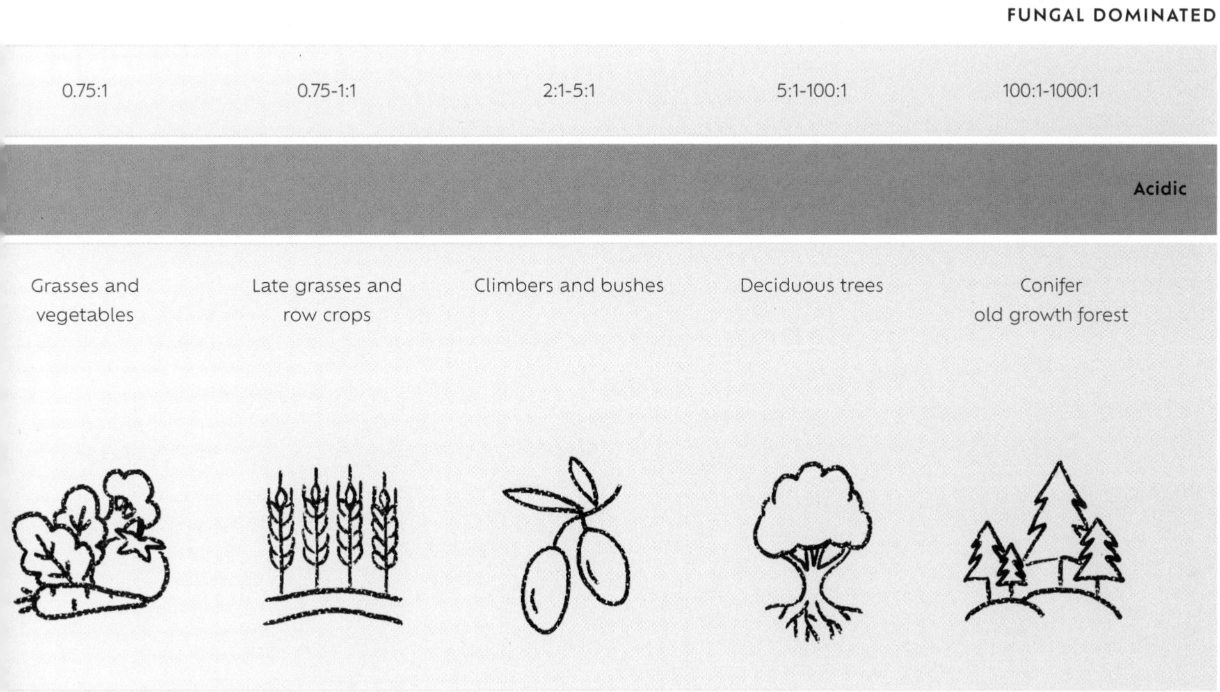

FUNGAL DOMINATED

0.75:1	0.75-1:1	2:1-5:1	5:1-100:1	100:1-1000:1
				Acidic
Grasses and vegetables	Late grasses and row crops	Climbers and bushes	Deciduous trees	Conifer old growth forest

Bacteria are the smallest and most hardy microbes in the soil and can survive under extreme and fluctuating soil conditions. They help with water dynamics, nutrient cycling and disease suppression. They can adapt to many different soil micro-environments (wet, dry, oxygenated or low oxygen). They alter the soil environment to benefit certain plant communities as soil conditions change.

Actinomycetes are classified as bacteria but are very similar to fungi and decompose recalcitrant (hard-to-decompose) organic compounds. Actinobacteria are responsible for the characteristically 'earthy' smell of freshly turned, healthy soil.

Fungi are microscopic cells that usually grow as long strands called hyphae, from a few cells to several metres in length. Hyphae push their way between soil particles, roots and rocks, and sometimes group into masses called mycelium or thick, cord-like rhizomorphs that look like roots. Fungal hyphae physically bind soil particles together, creating stable aggregates that help increase water infiltration and soil water-holding capacity. Along with bacteria, fungi are important decomposers, converting organic material into digestible forms for other organisms. Beneficial fungi can out-compete and destroy harmful fungi.

Protozoa are single-celled animals that feed primarily on bacteria, as well as other protozoa, soluble organic matter and sometimes fungi. They are larger than bacteria, at up to 2mm (1⁄10 in) in diameter. As they eat bacteria, protozoa release excess nitrogen near plant roots. Protozoa are classified into three groups: ciliates are the largest and propel themselves via hair-like cilia; amoebae move with temporary feet or 'pseudopods'; flagellates are the smallest of the protozoa and use whip-like flagella to propel themselves.

Nematodes are non-segmented worms typically 1mm (1⁄25 in) in length. Most nematodes are beneficial to soil ecosystems. Bacteria-feeding nematodes release plant-available nitrogen when they consume bacteria and also fix nitrogen, converting it into ammonium. Fungal-feeding nematodes have small narrow stylets (spears) in their stoma (mouth), which they use to puncture the cell walls of fungal hyphae and extract the cell fluid. This interaction releases plant-available nitrogen from fungal biomass. They also benefit plants by eating potentially pathogenic fungi, converting them to useful plant nutrients.

Predatory nematodes eat all types of nematodes and protozoa. They help to control populations, maintaining a healthy soil ecology.

Detrimental microorganisms tend to be present in anaerobic soils.

- Oomycetes are organisms that cause diseases such as seedling blight, damping-off, root rots, foliar blights and downy mildews.
- Ciliates eat amoebae and flagellates as well as bacteria.
- Root-feeding nematodes use stylets to puncture the thick cell wall of plant root cells and siphon off the insides, significantly damaging the plant. Many root-feeding nematodes carry pathogens. These can cause disease in plants, but in healthy soils they become good food sources and vehicles for spreading microbes and bacteria.

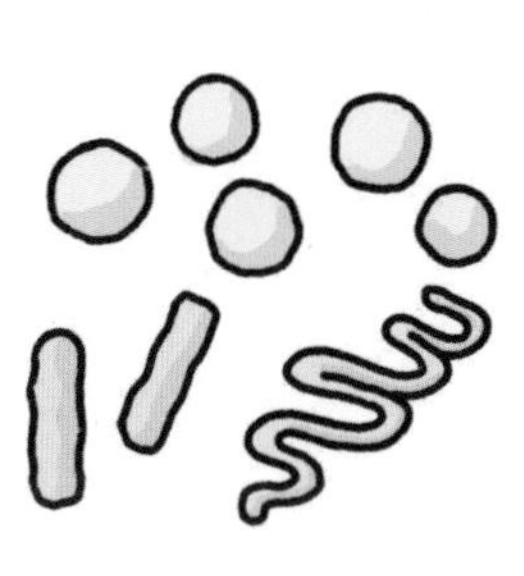

Bacteria

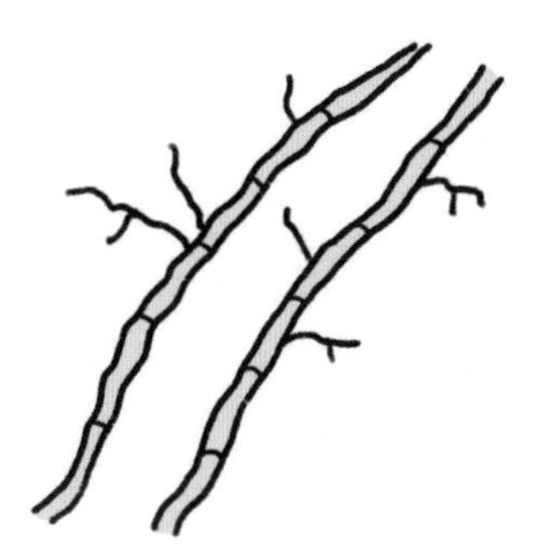

Fungal hyphae

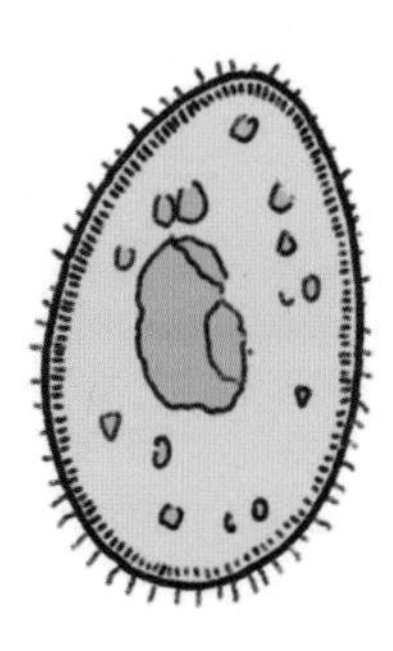

Protozoa

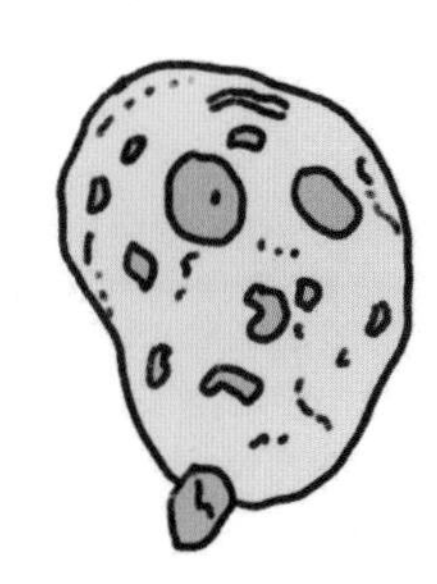

Amoeba

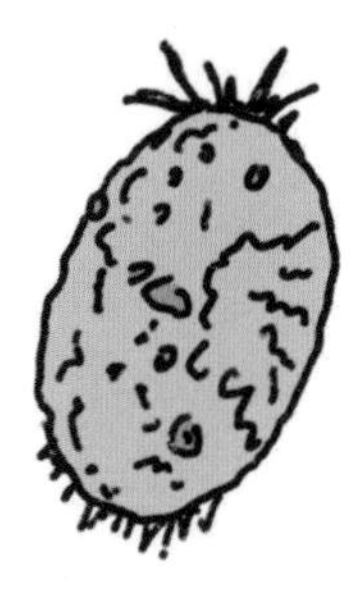

Ciliate

Fungi and bacteria eating nematode

Tardigrade

Root-eating nematode

HOW TO IMPROVE SOIL

> *How we manage our soil micro-herd is the key to regeneration.*
> *Nicole Masters*

1. **Know your mineral components**
- Is your soil mostly sand, silt or clay – they are all tiny 'pebbles' from parent rocks, which contain nutrients, in mineral components.
- Add living compost and organic matter to allow air, water and nutrients to pass between microbes and plants.
- Avoid compaction with an integrated layer of complete organic matter to protect the soil structure.
- Ensure there is a minimum 3 per cent organic matter to allow growth conditions.
- More is better. Life is made easier for plants and the growers if you have 5, 10 or 30 per cent organic matter.
- You could have organic matter alone – it's a great growing medium!

2. **Avoid toxic chemicals, which will kill the soil microbes and turn your soil into dirt**
- Avoid all 'cides' so no pesticides, herbicides or fungicides, including seed coatings.
- Avoid fertilisers that are made of salts – salt will kill your microbes too.

3. **Minimise digging, which slices up the soil structure and fungi**
- For a new bed or border, put down a layer of cardboard then compost and plant into that. Plants soon get their roots down into the soil below to access its nutrients and improve its structure.

4. **Plant a perennial layer of ground cover – a few centimetres (inches) of height is enough. Below ground, roots from the perennials will put exudates into the soil, keeping microbes fed and healthy even through most winters**
- Plant your bed or border through this layer. In the vegetable garden include perennial vegetables among annuals to maintain soil health. Save your own seed.

5. **Protect soil with mulch**
- Prevent raindrops destroying structure and causing compaction at least 10–15cm (4–6in) deep – more for a downpour.
- Avoid summer moisture evaporation bringing soluble nutrients to the bare surface, leaving a layer of salts, which damage plants.

6. **Be curious. If there is inconsistent growth, ask why**
- Is it compacted, waterlogged, over-fertilised or over-weeded?
- If one side of a border is

Layered planting will protect soil, build structure and grow healthier plants.

growing more than another, is there something buried underneath that's feeding that side or is there more compaction (or root-eaters) on the less happy side?

7. **Check your context – what are your goals?**
- Measure what is there, knowing the result you want (see What can we measure in our own gardens, page 196).
- How are you going to make the soil the best for what you want it to be? This does not always mean the richest possible. You can't grow healthy plants in bacteria-dominated soil as most of the nutrients end up inside the bacteria – this is a constipated soil.
- Some ecosystems like an early succession meadow will want a less rich soil (see Soil succession, page 48); think of the brownfield gardens on old building sites.

8. **Successional stage**
- Check the ideal fungal to bacterial ratio of the plants you want to grow: for example, weeds don't need fungi; broccoli wants some; grasslands would like a balance of fungus and bacteria; herbaceous borders then shrubs want more fungi; old growth forests want considerably more fungi.

9. **Get your soil tested – is it soil or dirt?**
- What are the beneficial organisms?
- Where do they fit in the soil food web?
- What are you missing, and how can you get the beneficial organisms to work for you?

10. **Once you have understood the biology you have, how can you improve it?**
- Add an inoculant from a local compost maker who has their soil tested as complete and pathogen-free. You don't need much to start the beneficial microbes growing in your own soil. You can add this direct, or make a compost extract, compost tea or indigenous microorganism (IMO) extract (explained below).
- If you have high 'weed' pressure, use biology to reduce the prevalence of plants you don't want.

11. **Ensure you are layering your planting to protect your soil, build structure and grow healthier plants**
- Plan protective plant material around any new planting. This can be annual or perennial – remember that a raindrop hitting soil can have the same impact as a tractor, deep down.

MAKING COMPOST, EXTRACTS AND TEA

The thermophilic method

There are many ways to make compost. This is a thermophilic method popularised by Dr Elaine Ingham and is similar to the method pioneered by Dr Johnson and Dr Su. It is also perfectly valid to have a 'normal' cold compost heap that you add to and leave for several years. The advantage of the thermophilic method is that it is quick and the heat kills weed seeds and pathogens, while keeping it aerobic means you avoid killing the good microbes and breeding anaerobic microbes, which will harm your plants.

You will need:

- Palette to raise the pile off the ground, to avoid waterlogging and to allow air flow
- Roll of stock wire to retain the compost – you can add weed membrane if conditions need it
- Staple gun and staples
- Galvanised wire
- Pipes with holes in and a long broomstick or similar to make holes in the compost to insert pipes
- 10 buckets the same size (or reuse fewer)
- Tarpaulin for covering the compost heap when needed and for turning the compost, every three days over fifteen days
- Soil thermometer – choose a galvanised steel one, 50cm (20in) or longer. (Aluminium tends to bend.) See online tutorials for how to recalibrate the thermometer once a year or so
- Rainwater (or tap-water left out to reach ambient temperature and with chloramine neutralised)
- Pitchfork or fork, to turn compost

Dry ingredients, in multiples of:

- 6 buckets (60 per cent) woodchip, brown leaves, shredded cardboard without ink
- 3 buckets (30 per cent) green waste from bokashi bin, grass clippings. Note: If you are saving green waste over time, spread it out so it can dry until you need it, rather than in a pile. If air cannot flow it will start to decompose anaerobically and turn into sludge
- 1 bucket (10 per cent) high-nitrogen ingredients such as alfalfa, phacelia, chicken poo, manure

Recycling waste into compost is one of the most satisfying and helpful things we can do.

MAKING A COMPOST BIN

Take the palette and position it in a cool, relatively shady spot in a temperate climate or in a barn or similar in extremely cold places. Form a wire frame just inside the palette and secure it with staples to the palette as a frame. Tie closed with wire, ensuring ease of reopening to turn the compost. Insert pipes with holes either before adding compost, or at the end if the pile is not too high and you can get the leverage from above.

The Johnson Su method uses perforated pipes to allow air to reach all parts of the compost heap to prevent it going anaerobic.

MAKING THE COMPOST

Make the compost

1. Assemble the ingredients, using the same size buckets so you get the right proportions.
2. Make sure the ingredients are about 50 per cent moisture when they go in (squeeze a handful tight in one hand, and if you can get one or two drops out that is moist enough). The buckets of brown material will usually need additional water, so fill halfway with brown ingredients and top up with rainwater from a butt. If you need to use tap-water add some humic acid or vitamin C preparation to neutralise the chloramine first.
3. Put the ingredients into the compost bin; the order is not important.
4. Using the end of a rake or broom handle, make holes into the compost heap and insert the pipes with holes, if you haven't already, making sure they are spaced throughout 30cm (12in) apart so that air can reach all parts of the compost heap to prevent it going anaerobic. Cover with a tarpaulin in extreme heat, cold or wet; if you can drape this over a central post you will allow any condensation to fall back into the heap, retaining moisture.

The aim of turning compost is to allow each section time in the middle to get hot enough to kill pathogens and seeds without staying too long and becoming anaerobic.

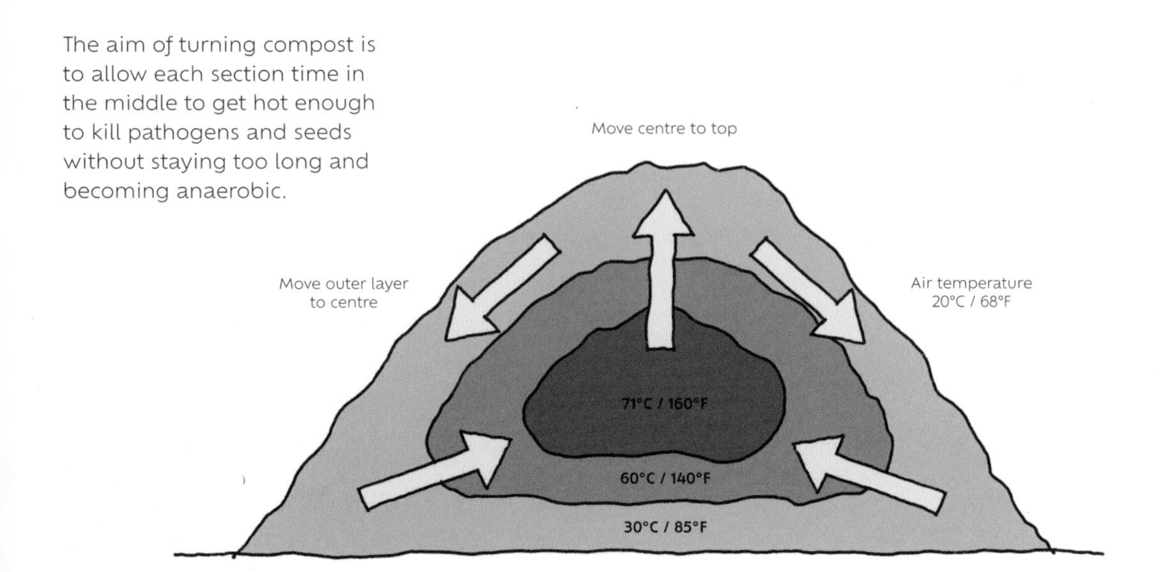

Ensure all compost is turned, moving outer layers to centre where temperatures are hot enough to kill pathogens. Repeat 3–5 times.

5. Using the soil thermometer check the temperature in the centre of the heap after 24 hours and again after 36 hours and 48 hours. If it gets above 55°C (131°F) within 36 hours, add a sprinkling of ambient temperature water from a water butt. After 48 hours if the temperature has stayed between 35°C (95°F) and 55°C (131°F), it is time to turn the heap.
6. Use the pitchfork to take the top third of the compost off the pile and put on the tarpaulin (pile 1). Then remove the 'hot' middle. Place this in a pile (pile 2). Finally take off the bottom and place in a third pile (pile 3). Replace the compost in the frame as follows: put pile 1 at the bottom; pile 3 goes in the middle to get a chance to get hot; and pile 2 goes at the top. If there are any lumps break them up so that the ingredients all get a chance to break down evenly. Replace the pipes and the tarpaulin if using.

Repeat steps 5 and 6 after another 24 to 36 hours, turning after 24 hours if the compost is getting too hot or leaving it a little longer if the thermometer stays happily in the green zone.

After five turns over fifteen days your compost should become less active and stay happily in the green or yellow areas of the dial on the thermometer. The weed seeds and pathogens should have been killed, and your compost is ready to go into the maturing phase. During this stage it is worth looking at it under a microscope periodically to see if you have the whole soil food web present in the biome, or if it would be beneficial to add anything else before using it as a soil additive, foliar spray, seed preparation or mulch.

If you don't have the whole soil food web, you might add an inoculant from a local compost maker who has tested soil that is complete and pathogen-free. You don't need much to start the beneficial microbes growing in your own pile. You can add this direct as compost or make a compost extract.

Below: Using the soil thermometer, check the temperature in the centre of the heap after twenty-four hours and again after thirty-six hours and forty-eight hours.

Opposite: Use buckets to measure proportions of 60 per cent brown and 30 per cent green waste, plus 10 per cent high nitrogen 'party food' to activate decomposition.

MAKING A COMPOST EXTRACT OR TEA

It is very simple to make a compost extract, which can be used to soak seeds to add beneficial microbes before sowing, or to add to chloramine-treated water, to neutralise the chemicals in mains water, and add the important nutrient-dispersing workforce before it is poured on to compost or around plants.

1. Put a handful of complete compost in a hessian bag and gently massage it in a 10 litre (2 gallon) bucket of water for about 60 seconds. The aim is to extract the microbes from the soil and suspend them in the water, so do this until the water turns the deep brown colour of 70 per cent chocolate.
2. To make the extract more effective you can turn it into a tea by adding aeration, either by whisking by hand or, as in the image left, using a handheld drill with a plaster mixer bit. The aim is to get a vortex going for about a minute in each direction to add this energetic force as well as thoroughly extract the microbes into the water. This tea should be consumed within 48 hours and can be used in very small amounts direct on to foliage. Simply take a switch or bunch of leaves and flick it out of the bucket on to the plants. For larger quantities a backpack sprayer can be adapted by adding a 20mm (¾in) nozzle to avoid killing the microbes through over-compression. For even larger quantities it is worth investing in a bubble brewer, which will add oxygen over a 24- to 48-hour period.
3. If you have highly bacterial soil you can make this into a protist tea by soaking a handful of dried organic alfafa in a 10 litre (2 gallon) bucket for three days before bubbling it with the compost extract to make the tea. To make the tea more fungal, you can add a very small amount of fish hydrolysate and use this to spray on to plants as a foliar boost to their health and resilience.

MAKING INDIGENOUS MICROORGANISMS (IMOS)

You will need:

- Dried rice
- Saucepan
- Water to boil the rice in
- Wooden box, such as a wine box
- Newspaper or brown paper and string
- Chicken wire (optional)
- Spade

If you have depleted soil in your garden, you can harvest indigenous microorganisms from a local woodland.

1. Put the water and rice in the saucepan and cook until tender.
2. Then put about a 7–10cm (3–4in) layer of the cooked rice on the bottom of the wooden box. Do not compact the rice as you want to allow air and microbes to move around.
3. Wrap the box in paper and secure with string to allow air in and out – but not animals. (I added chicken wire to keep out foxes and rats, etc.)
4. Bury the box in a woodland where leaves fall and are decomposed by fungi. Leave out for 7–10 days in moist conditions and then dig it up. The time needed will depend on outside temperature (less time will be required in warm weather).
5. When you undo the box you should see a white fungal layer growing on the top of the cooked rice. This can be added to your compost heap to increase its fungal content and microbial activity, to make a more complete food web and speed up the nutrient cycle. For ways to store this and add more amendments see resources section.

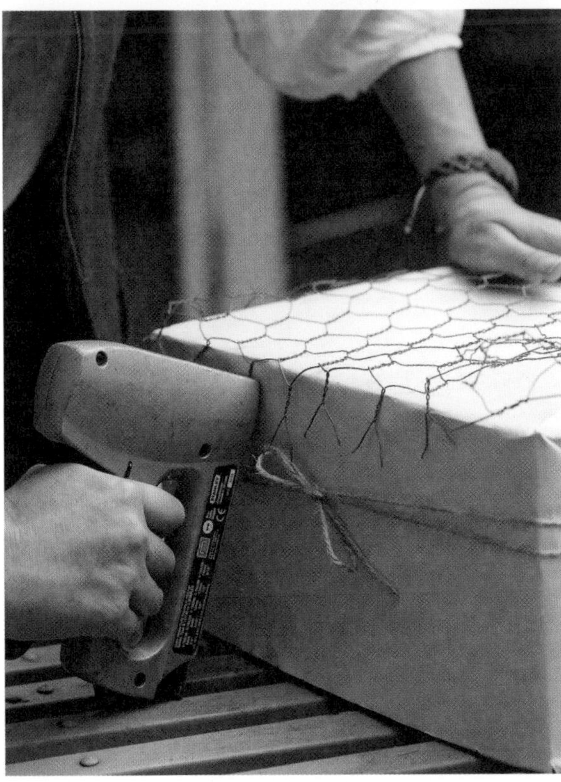

WEEDS

> *Weeds should be reclassified as doctors of the soil.*
> Nicole Masters

Weeds are our indicator plants. They tell us what is going on under the soil. Some colonise bare ground, like scarlet pimpernel (*Anagallis arvensis*) and chickweed (*Stellaria media*), or hold together early succession soils with creeping fibrous roots (see Soil succession, page 48). Others remediate compacted soil by breaking it up with long taproots, like docks (*Rumex*) and thistles (*Carduus, Cirsium*). They bring up minerals, nutrients and water from deep below, creating water channels below the ground and feeding wildlife above it. As they break down, they add organic matter and the mined minerals return to the topsoil.

Before you dig out a weed, consider what it is telling you and how you might benefit from it, either in a tea for yourself, in compost or tea for your plants, or as forage. Is it there to remediate compaction, perhaps for excess toxins or lack of minerals? Look up what it accumulates and what it is indicating.

Before you dig out a weed, consider what it is telling you and how you might benefit from it, either in a tea for yourself, in compost or tea for your plants, or as forage.

Nettles (*Urtica*) and ground elder (*Aegopodium*) can both indicate low calcium, for example, so you could try spraying any left-over milk, diluted with rainwater, or crushed baked eggshells. Adding this calcium source may help rebalance your soil enough so the weeds no longer get the signal to grow.

If after this you want to reduce them, try biology. For too many nettles or ground elder, flatten them by stamping or rolling and apply a thick layer of woodchip. This will help suppress their growth by reducing sunlight to leaves and new shoots, allow them to break down and return the minerals they have accumulated to the soil and – as the woodchip breaks down – the higher fungal dominance in the soil will make it less conducive to their growth.

See the table opposite for some common weeds, what they are telling us and what we can use them for.

COMMON 'WEEDS': WHAT THEY ARE TELLING US AND THEIR USES

Weed	Latin Name	Soil Indication	Benefits	Medicinal Properties and Uses for Humans
Dandelion	*Taraxacum officinale*	Compacted soil, low calcium and pH imbalance	Draws up nutrients, improves soil structure, edible (leaves and root). Adds calcium and silica to compost. Earthworms use root channels.	Diuretic, liver support, rich in vitamins and minerals, used in traditional medicine for various ailments
Nettle	*Urtica dioica*	High nitrogen, fertile soil	Dynamic accumulator, improves soil fertility, edible (young tops and seeds). More calcium, sulphur, boron and phosphate than alfalfa.	Nutrient-rich, anti-inflammatory, traditionally used for arthritis, allergies, and as a tonic. Used in Biodynamic BD 504 for iron.
Plantain	*Plantago major*	Compacted soil; low fertility	Breaks up compaction, nutrient accumulator, medicinal properties, seeds source of protein	Anti-inflammatory, wound healing, leaves sooth skin irritations and insect bites
Shepherd's purse	*Capsella bursa-pastoris*	Disturbed, nutrient-rich soil	Soil stabiliser, attracts beneficial insects, removes salt from the soil	Traditionally used to reduce bleeding, menstrual issues and as a diuretic
Chickweed	*Stellaria media*	Rich, fertile soil	Ground cover, attracts pollinators, edible (young shoots)	Used for skin conditions, as a mild diuretic and as a nutritive herb
Bindweed	*Convolvulus arvensis*	Poorly drained, compacted soil	Erosion prevention. (Can be killed by root secretions of tagetes and dahlias.)	Not commonly used medicinally
Horsetail	*Equisetum arvense*	Poorly drained, high acidity	Bioaccumulator for silica. Natural fungicide used in BD 508	Diuretic, anti-inflammatory, traditionally used for wound healing and bone health
Ragwort	*Jacobaea vulgaris* (previously *Senecio jacobaea*)	Poorly drained, acidic soil	Attracts beneficial insects, can be toxic to livestock	Not used in herbal medicine due to toxicity
Couch grass	*Elymus repens*	Poorly drained, compacted soil	Erosion control, may prevent soil degradation	Traditionally used for urinary tract infections
Spear thistle	*Cirsium vulgare*	Poor fertility, well-drained soil	Attracts pollinators, wildlife habitat.	Some species have historical medicinal uses for liver support, but caution is advised due to potential toxicity

WATER: HOW DOES THE WATER CYCLE WORK?

We are a wave appearing on the surface of the ocean.
Thích Nhất Hạnh

Our bodies are 70 per cent water, and we need to drink to hydrate every day to thrive, yet water is finite on the Earth, so with 8 billion people on the planet how is water sustained?

Water lives in the oceans, lakes, rivers, plants and living creatures including us. It continuously cycles through evaporation, condensation and precipitation (rain) in a miraculous perpetual loop (see diagram overleaf). The water in your tea has likely been drunk millions of times before. It holds within it a molecular memory – a part of us connected to our ancestors, enemies or even dinosaurs.

There are few feelings as freeing as swimming in clean open water (naked if appropriate). It's a return to our innate connection with nature. The endorphins released during cold-water swimming have proven health benefits, enhancing mental focus and potentially increasing life expectancy. If you're new to it, start in summer when water is warmer, limit a colder dip to a minute or two, enter slowly and use controlled breathing and core warming to keep your body safe and your mind keen to return, and of course only swim where clean and safe to do so.

As well as reconnecting us with nature, the resurgence in wild swimming has highlighted the plight of our waterways. Movements like We Swim Wild and Surfers against Sewage have raised awareness of our water quality issues and provide information on water quality and safety, so check their apps when choosing where to swim. Direct sewage discharge into

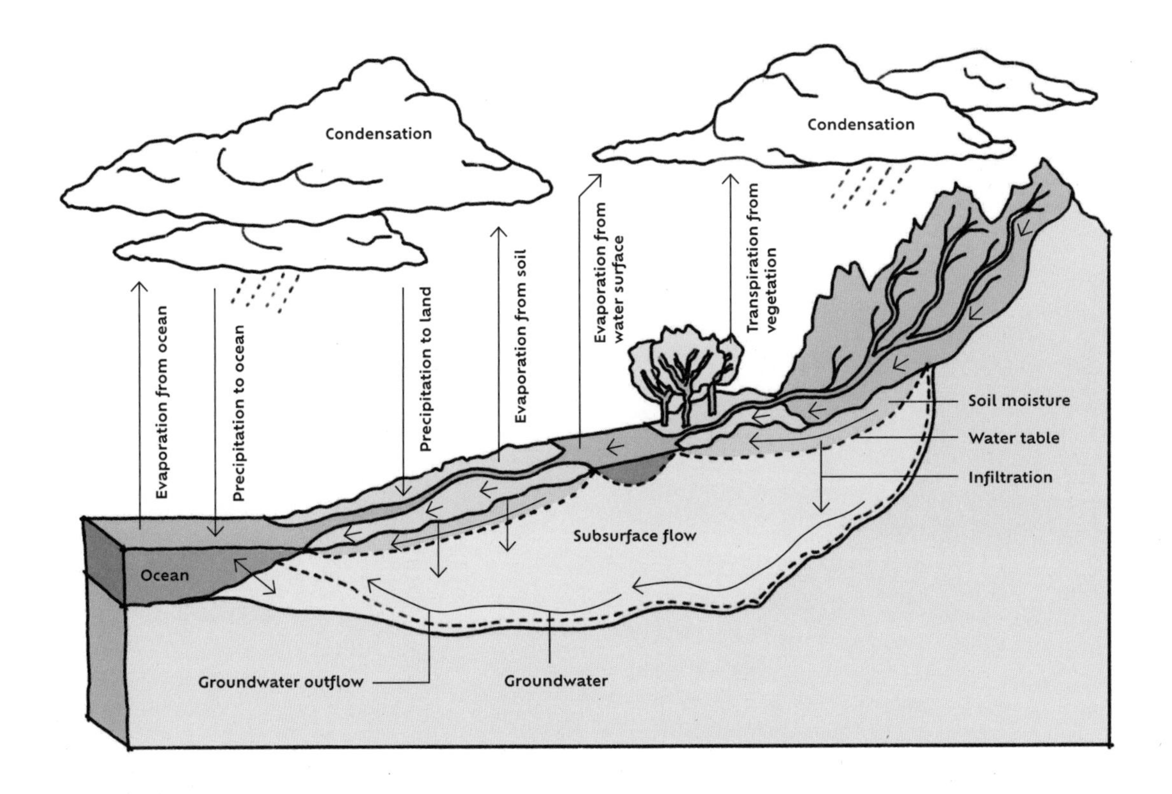

Above: Water continuously cycles through evaporation, condensation and precipitation (rain) in a miraculous perpetual loop.

Opposite: The water in your tea has likely been drunk millions of times before. It holds within it a molecular memory – a part of us connected to our ancestors, enemies or even dinosaurs.

our rivers is harming our water and wildlife. The eutrophication or algal bloom this causes can also cause blue baby syndrome and death.

Pollution also comes from faeces from livestock grazing too close to water, from slurry on fields and run-off from industrial or factory farms above watercourses. This is made worse when the surrounding soil cannot absorb water. On top of this, chemical fertilisers add nitrogen and phosphates to the water, and several industries including some fisheries still use the sea as an open sewer.

To secure clean water for future generations, we can campaign for changes in water management by water companies, safer agricultural practices and planning and industrial regulations. We can play a crucial role in improving water quality by advocating for our rivers to be given rights and protection, by working to make swimming in our lakes and seas safe for all, reporting pollution and responsibly managing our own land to be the guardians of the water that falls there.

Wider movements are rewiggling rivers and restoring watersheds and floodplains, and slowing the flow with ecosystem engineering like beavers. There is an energetic shift in the way we view our water; water companies are being challenged and we are realising that water should not be viewed merely as a means to flush away waste. We are beginning to recognise it for the life force it is.

On a personal level all choices have an impact. You can support eco-friendly companies, reduce use of plastics and chemicals, and ask questions as a consumer and water company customer. Opt for natural fibre clothes that can be repaired and eventually composted to avoid the microplastics in synthetic clothes like fleeces getting into the water. Use a 'guppyfriend' bag to wash synthetic clothes as it will capture microplastics to dispose of in land fill, slowing down their journey into the water and into our guts. A compost loo saves on mains drinking water and adds to soil health; when composted with sawdust, ash and woodchip, it makes valuable fertiliser for non-food plants.

Here is a challenge: how much drinking-grade water do you use per day? Studies recommend a maximum of 110 litres (24 gallons) per person per day, to keep within planetary boundaries. Changing just a few habits or using collected

CONSERVATIVE DOMESTIC WATER USE IN A WEEK (PER PERSON)

Activity	Frequency per Week	Water Use (litres)	
Clothes washing	1	50	
Dish washing – modern dishwasher on eco-setting	2	22	
Daily showers – electric shower	7	250	3 x 10 mins and 4 x 5 mins = 50 mins
Brushing teeth (twice a day, tap not running for the entire brush)	14	17.5	10 secs of tap running per brush = 2 mins 20 secs
Flushing the toilet – modern cistern (six times a day)	42	210	
Washing hands for 30 seconds after the toilet (tap running)	42	157.5	42 x 30 secs = 21 mins
Daily water intake (2 litres)	7	14	
Cooking (filling a 1.5 litre saucepan once a day)	7	10.5	
		731.5	Litres per week
		104.5	Litres per day

	Per Day (litres)	Per Week (litres)
Environment Agency target per person	110	770
Water left over for gardens and other activities	5.5	38.5

This leaves enough water for just 2 mins 18 secs of hose pipe use per week!

Opposite: Water is a vital cooling agent. Just sit by a pond or the sea on a hot summer's day to feel the difference it makes.

Above: Here is a challenge. How much drinking-grade water do you use per day? Studies recommend a maximum of 110 litres (24 gallons) per person per day, to keep within planetary boundaries. Changing just a few habits can have a big impact.

rainwater instead of mains can have a big impact. See the table above showing water use by some common activities. What small changes could use less mains water, or avoid onward contamination from chemicals, hormones, antibiotics and microplastics?

Water is a vital cooling agent. Just sit by a pond or the sea on a hot summer's day to feel the difference it makes. On a global scale water vapour buffers atmospheric global temperature, keeping us cooler and more comfortable.

Trees, plant cover and soils are crucial in impacting rainfall levels, as rain that falls is recycled through plants by transpiration and moisture is held in healthy soils. Simply put, if we plant ecosystems and create healthy soils, it rains more regularly and the Earth remains cooler. If we cut down our forests and woodlands or leave large areas of soil bare, the reduced cover quickly increases the dust-bowl effect. Deep roots bring water up in times of drought, and transpiration feeds the local water loop. The way we manage our land has a big impact on cooling the planet, and this includes our own gardens.

WETLAND REGENERATION: WHAT WE CAN LEARN FROM LARGE-SCALE PROJECTS

> *If you build it, he will come.*
> *Field of Dreams*

As the increase in flooding in recent decades has shown, our rivers have become too straight and too fast, the banks are too steep and water runs off fields taking soil and nitrates to the sea, creating a cycle of destruction. Re-meandering our rivers to slow the flow and regenerating the banks and soils in fields to hold water for longer is vital to our collective health. Trees along banks cast shade for breeding fish and provide homes and perches for birds. Dropped leaves and branches add biomass to the water to help create healthy ecosystems. In time these become permeable dams, trapping sediment and pollutants.

If you want to re-meander, to remove a weir or construct a dam on a UK river you will need a licence, but it is possible, and many

Wetlands' unparalleled capacity for water retention is shown in this photo.

authorities are open to discussing improvements. (Details of where to get help can be found on page 230.) In the UK several projects in recent years have focused on slowing down rivers and restoring wetlands. It is one of our most valuable ecosystems, able to lock up the most carbon of any habitat, and accounting for 10 per cent of species despite covering only 1 per cent of land on Earth.

There are two main ways to do this. The first is being pioneered by the National Trust's Holnicote Estate in Devon, where land drains have been broken, wide scrapes dug and the old narrow river channel filled in, to allow water to infiltrate the historic floodplain naturally. This will raise the water table and create multiple and diverse pockets of habitat as the water finds its natural course through the plain. To help create habitats, whole trees have been laid like horse jumps as obstructions, waiting to collect debris like beaver dams. The project is called 'Stage 0' as it returns the floodplain to its original stage before humans began to drain it for agriculture. Dr Richard Brazier and Dr Alan Puttock at the University of Exeter have modelled predicted flow and optimum ground levels, and baseline surveys are monitoring the species arriving, the level of water cleaning and flood prevention taking place.

Another project being guided by the University of Exeter with others is at the tiny headwater stream on the river Tamar, where a 900-metre (1,000-yard) fence created a trial beaver enclosure in 2011. It monitors the effect of these animals on the landscape and whether they prevent encroaching scrub from overwhelming the unmanaged grassland. As the image at the beginning of this section shows (page 68), the difference in landscape character and biodiversity is astounding. The beavers created highly complex, ever-changing mosaics of rich micro habitats and brought back rare water beetles, damselflies, dragonflies, frogs, newts and toads, plus kingfishers, water rail, roosting snipe, grey herons, willow tits, grasshopper warblers, black caps and willow warblers. As well as the excellent biodiversity and carbon capture, the ecosystem service impact really showed in times of recent drought, in stark contrast to the surrounding land, which was parched and suffering. This 3-hectare (7½-acre) site now holds around 1 million litres (220,000 gallons) of water. As well as drought resilience it also slows flow during storm events, with results showing stormflow reductions of up to 30 per cent downstream.

At Spains Hall Estate in Essex, regenerative land steward Archie Ruggles-Brise has worked with local water companies and authorities to create wetland habitat and clean water with two beaver enclosures. By holding water in high-rainfall events, the new wetlands have reduced the historic flooding that their local village suffered, as well as increasing tourism in this rural area.

Beavers are herbivores; they don't eat fish. They eat lots of willow (*Salix*), for which they were hunted to extinction in England, as the castoreum produced in their scent glands contains the medicinal silicic acid from willow (the main component in aspirin). They also eat herbaceous plants like the rampant watercourse colonisers Himalayan balsam (*Impatiens glandulifera*) and hemlock water dropwort (*Oenanthe crocata*). They can fell most types of trees with their iron-toughened teeth, to create habitat and to reach the succulent new growth on the upper branches. Once we understand how efficient and effective beavers are at creating useful and beautiful habitat, the questions become: why don't we immediately introduce them everywhere? And why are farmers so wary of them?

It's important to understand all concerns in this coexistence vision, and to know that most smaller farmers survive on tiny profit margins or are in debt to large agricultural input corporations, with rent or mortgages to pay on their land. Prey to uncontrollable weather and changing government ideology and paperwork, beavers can seem just another unpredictable factor in farmers' volatile futures. The facts are that beavers will take down some unprotected trees, and they will eat unprotected carrot or potato crops near their homes if available. They may dam in inconvenient places or even burrow under roads. We can, however, learn to live with them and other future ecosystem restorers by looking to places that have coexisted with them during the centuries we lost to extinction. In Bavaria, Germany, each area has a volunteer beaver warden who liaises with farmers and landowners, educating, negotiating and, if needed, translocating or even euthanising their 25,000 beavers. Farmers are paid by the taxed community to leave a 9-metre (10-yard) riparian strip along water banks to prevent crop damage. Many find this a useful income compared to the tight margins on high-input crops. Key trees are protected with wire netting, and important infrastructure is constructed to be beaver-proof, with simple rebar construction on banks to prevent burrowing.

In the UK the Beaver Trust is helping to educate and increase dialogue with landowners and stakeholders, and Natural England,

We are all an intrinsic part of nature and in our own gardens there are usually no beavers, nor bison, wolf, elk, wildcat or wild boar. So, along with the bees, we are the remaining prime ecosystem engineers.

NatureScot and Natural Resources Wales are all working to advise ministers and influence nature-positive policy. All of us can help by being informed, by amplifying the messages and by supporting our wildlife trusts with money and time.

We are all an intrinsic part of nature and in our own gardens there are usually no beavers, nor bison, wolf, elk, wildcat or wild boar, so along with the bees we are the remaining prime ecosystem engineers. We reap the benefits of the ecosystem we create, through planting like a jay or squirrel or through disturbance, whether coppicing to harvest and maintain woodland or pruning roses like a browsing herbivore. Imagine you are rootling like a wild boar when weeding, rejuvenating and creating a little space for new growth, or building mini dams like a beaver to make water slow down and stay, bringing extra life to your garden.

It is up to us to create a regenerative rather than destructive garden. Wouldn't it be great if homeowners were incentivised to provide ecosystem services in our gardens. Council tax could be reduced for a Five Star Rated Garden that locks up carbon, reduces temperatures in summer, cleans water and improves soil health. Until then we can enjoy just being part of the solution.

WATER IN OUR GARDENS

> *You are not a drop in the ocean.*
> *You are the entire ocean in a drop.*
> *Rumi*

Taking care of water in our gardens can make a big difference. The best thing to do is to catch it in your soil. Healthy soil will hold up to seven times its weight in water and allow it to filter slowly down to the aquifers. Even heavy clay when it has lots of humus and good texture created by microbes will be spongy, allowing us to walk across a garden after rain without sinking. A healthy soil is up to 80 per cent air pockets, while a compacted soil may only have 2 per cent space left for water and the nutrients it carries to plant roots. A soil infiltration test will tell you how well your soil is holding water (see What can we measure in our own gardens?, page 196).

As well as improving soil, catch rainwater to use on your garden. This takes pressure off the mains and avoids the chlorine and chloramine added to tap-water, which kills soil microbes. If you do use mains water on your garden, leave it to stand in a water butt or watering can for 30 minutes to allow chlorine to evaporate, add a bit of humic acid to the water to mitigate the chloramine, or add a living compost or compost tea to your soil after watering to help replenish those crucial microbes (see Making compost, extracts and tea, page 56).

Calculate how much rainwater you can harvest from your roof (see What can we measure in our own gardens?, page 196) and invest in water butts, or a rainwater harvester or pond in a bigger garden. These methods reduce pressure on the water and sewerage systems, which are struggling to keep up with old infrastructure and increasing populations. Urban drainage systems were designed to take rain and waste into sewers as fast as possible. Rivers were culverted and front gardens paved over. Some 50 per cent of London is now hardscape (and 60 per cent of New York City).

Sustainable drainage systems and green spaces would hugely benefit our cities. In our own gardens we can create as much green surface as possible to slow the flow of water and lessen its impact on flooding and nutrient loss. This can include green roofs, planted gravel drives, climbers on walls and hedges, or even a fence covered in climbers (edible if possible).

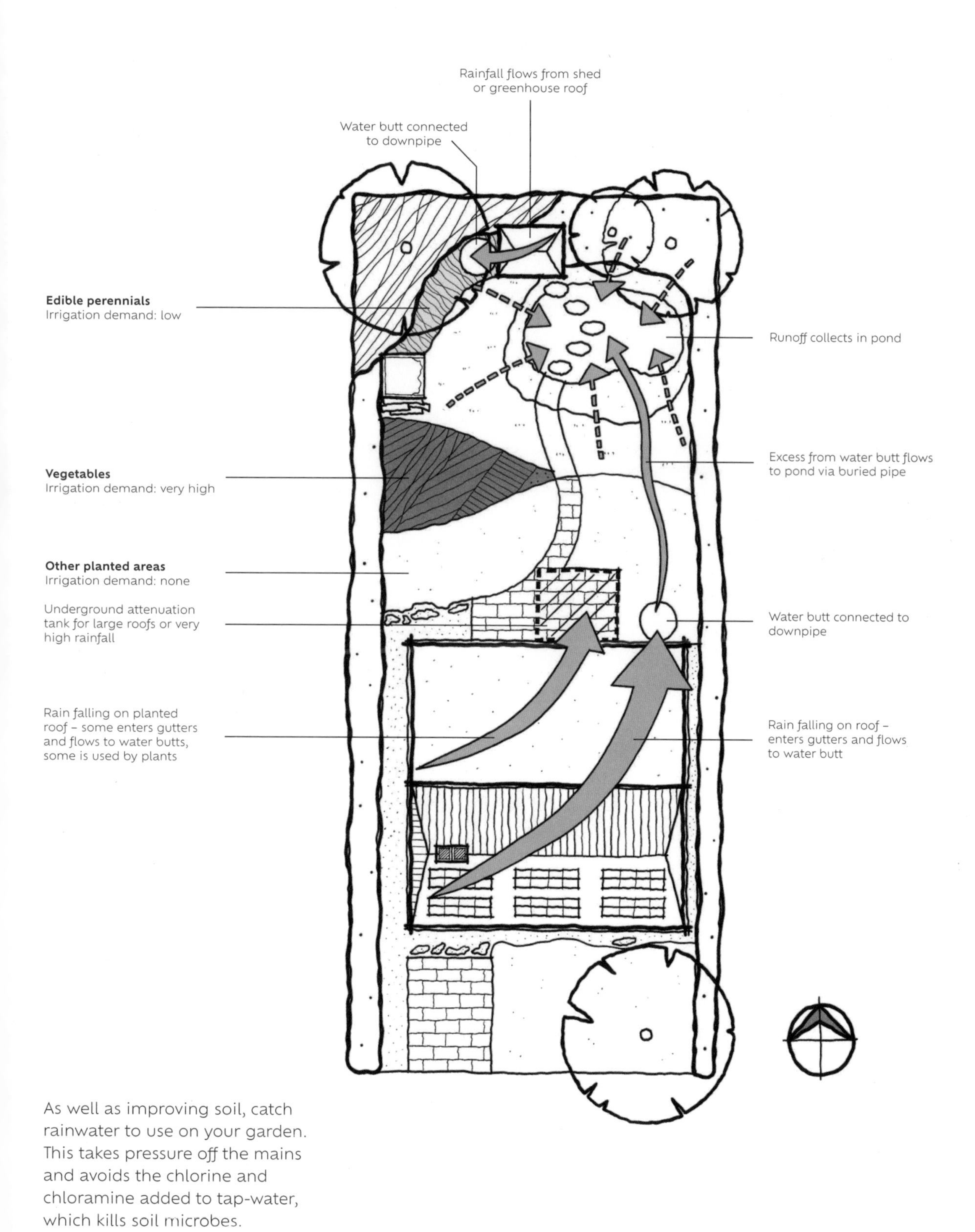

As well as improving soil, catch rainwater to use on your garden. This takes pressure off the mains and avoids the chlorine and chloramine added to tap-water, which kills soil microbes.

Above: To reduce water usage plant for local conditions, create shade, cover soil with resilient plants, and feed your soil with microbes and mulch.

Opposite: The edges of ecosystems are where maximum biodiversity thrives, so wiggle your pond edges to make them longer, add swales and ditches and allow dew ponds and puddles to add to the magic.

Introduce a pond to revolutionise your ecosystem and improve the microclimate. This will attract creatures, retain water and emit cooling mist as it evaporates. A puddled or clay-lined pond will help increase water availability to plants around it (see page 83). Even a tiny pond, water bath or rain garden will add life to your garden, while a shallow dish or bucket of water can be a lifesaver for creatures in times of drought.

I am often asked how to prevent a garden pond from going green. The answer is in the biology. The green is algae, which grows fast in unbalanced ecosystems exacerbated by nitrates and warmth. Shallow ponds with black, sun-absorbing liners and few plants are a perfect breeding place for bacteria, algae and mosquito larvae, but less good for the beneficial organisms that would balance them out. The solution is to ensure some shade, adequate depth and plenty of varied surface areas. You can mimic the rich ecosystem of a shipwreck on a coral reef with different levels, gravel, stones, rocks, bricks and a sunken structure or two. Position oxygenators and floating-leaved plants at the bottom and plant marginals at the edges. The edges of ecosystems are where maximum biodiversity thrives, so wiggle your pond edges to make them longer, add swales and ditches and allow dew ponds and puddles to add to the magic.

To reduce water usage and the impact of using mains water for irrigation, plant for local conditions, create shade, cover soil with resilient plants, and feed your soil with microbes and mulch. Allow a dew pond to form in a wet spot, and plant water-thirsty willow (*Salix*) to coppice for woodchip and weaving. Transition from a labour- and water-intensive annual vegetable garden to a mixed permaculture garden; with mostly perennial plants that grow in self-supporting layers with some annuals between, soil is covered, and soil microbes and mycorrhizal networks kept intact.

If you're planting new trees, start as small as practicable, mulch with woodchip or wool and use irrigation bags to establish larger root balls. If space is limited opt for Mediterranean plants, which can thrive in gravel.

A CLAY-PUDDLED POND

> *I come into the presence of still water.*
> *And I feel above me the day-blind stars*
> *waiting with their light. For a time*
> *I rest in the grace of the world, and am free.*
> *Wendell Berry*

It used to be the norm to have a pond in most gardens from which to collect rainwater, wash the wheels of the horse cart or cool a bottle for a summer picnic. Many of these havens have vanished, yet if your garden has the space, a pond is the most impactful way to enrich your ecosystem.

The cooling effect of water in summer and the warming effect of evaporation in winter will make the garden more temperate. Bats, birds, amphibians and invertebrates all rely on water to drink, and newts, frogs, toads and insects like dragonflies and waterboatmen need water to reproduce. If you avoid fish in your pond you will have much greater diversity of other species. If you do have fish, allowing natural predators like herons will help prevent them over-dominating the other life in the pond.

Traditionally, ponds were clay-lined, created by 'puddled in' clay layers compacted by livestock hooves – a method both ecologically friendly and easily recyclable compared to a plastic or rubber liner. If part of your garden grows sedge (*Carex*) or visibly sits wet after rain, that's a good spot to dig a trial pit. If you find a clay layer at the desired depth, that is the simplest route to a successful pond.

If you don't have a naturally wet area, choose the lowest point in the garden. This spot will naturally collect water and could potentially be filled through a swale connected to downpipes or water butts. Before digging, locate tree roots and underground services to avoid both. Hitting a gas main is no laughing matter.

The ideal time to make a clay-puddled pond is autumn, outside of newt breeding season and when rains are expected. If it does not rain, the clay may dry out and crack and you'll need to do it again just before rain.

To shape the pond, create a gradual slope or several shallow shelves from edges to centre, enabling small animals to enter and escape, and make the sides wiggle a bit to allow for more edge and maximum biodiversity in this important 'draw-

A clay-puddled dew pond, to catch water and hold it for a while.

When designing pond planting, consider three zonal depths – marginal; second tier; and deep floaters or sinking oxygenators – and choose plants that thrive at each.

down zone', which may dry out in summer. Compact the clay carefully, either with a mini digger and sheepfoot roller, or a herd of cows then sheep, or a makeshift 'herd' of adults followed by children. Run the larger then smaller 'hooves' or tracks back and forth over at least 30cm (12in) of wetted clay in all four directions to ensure the base is watertight.

Check for leaks after rain (put a measure in to mark the water level and come back regularly to check) and increase the thickness of clay if leaks occur. It is possible to buy loose quarried puddling clay or bentonite to infill cracks. This comes by the tonne bag or lorry load, and you should allow about 0.5 tonnes per square metre (0.05 tonnes per square foot) at minimum 30cm (12in) depth.

If your pond is on chalk or sand you will need a thicker layer of clay, up to 1m (3ft), or you may opt for a bentonite liner, which is a roll of clay sewn between two geosynthetic layers. The ecosystem benefits will still be huge; however, the geosynthetic layers will be adding to microplastic pollution in the long term. (See Appendix 1: Comparison of pond liners, page 228.)

Once you have the shaped the pond, cover its sides with 30cm (12in) of soil and pebbles, and position some rocks or large pieces of rubble around the edges to provide crevices and shelter

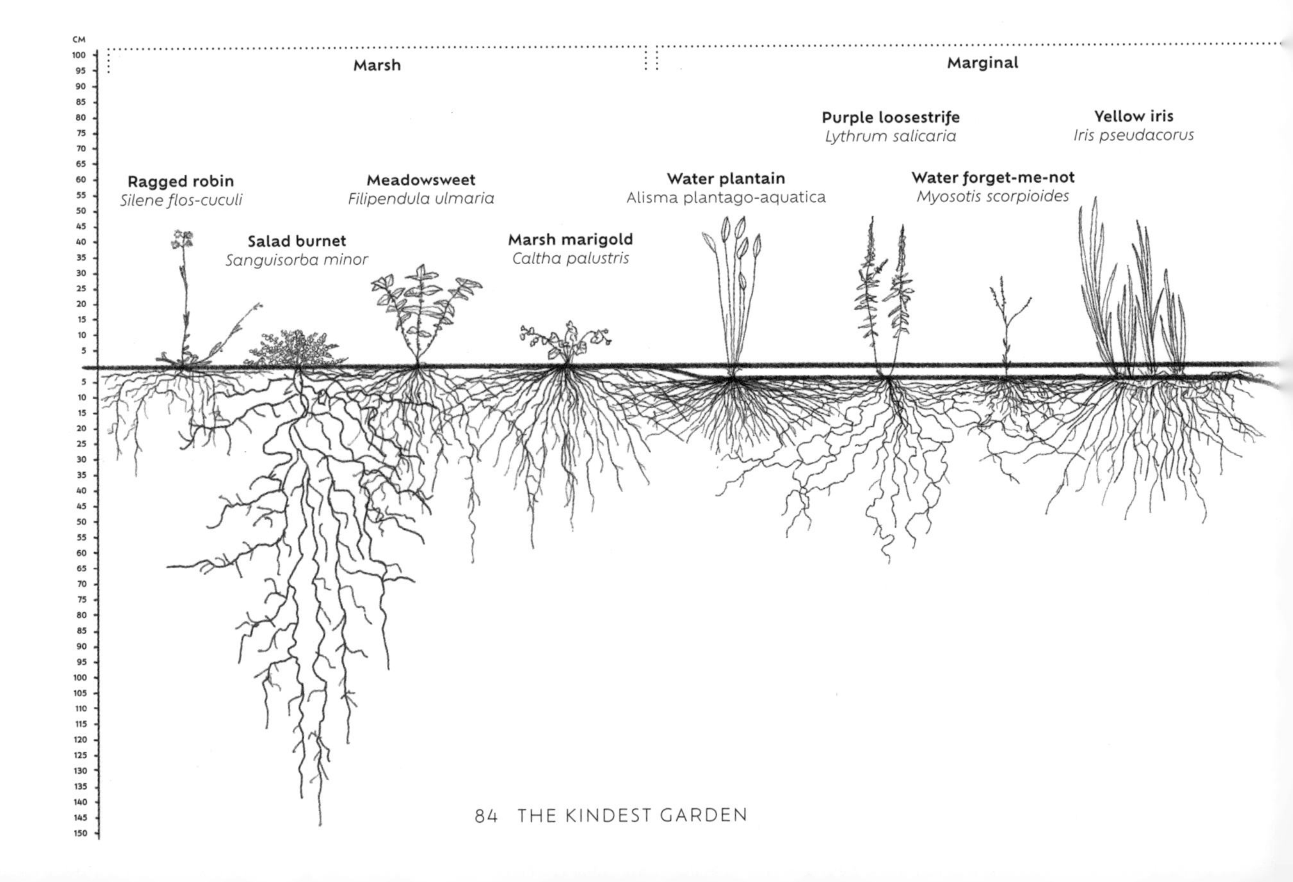

for small animals. Any other items you add will increase the surface area for beneficial bacteria and plants to colonise. Aim for 50 per cent shading to keep the water cool, through nearby shrubs and trees, floating plants and even a wooden boardwalk or jetty to allow you to lie on your tummy and get close to the action.

When designing the planting, consider three zonal depths – marginal, emergents and aquatics or sinking oxygenators – and choose plants that thrive at each. Use a hessian roll filled with soil to establish edge plants. Lay a long sausage of soil mixed with clay on the hessian and sew up the ends, then cut crosses, to plant marginals at 30cm (12in) centres. Weigh this down with bricks or wedge it under rocks. These plants will slowly colonise the pond and get their roots down as the hessian decomposes. Deep-water plants like waterlilies (*Nymphaea*) can be wrapped in hessian and tied to a brick to colonise at 1m (3ft) depth.

If you have a large enough pond with an area of 3×3m (10×10ft), which you can guarantee will always have at least a 1.5m (5ft) depth of water, consider water source heat coils. They will reduce the temperature of the pond by up to 1 degree and transfer a small amount of heat from the pond to a heat store, which can be used in the house as part of a renewable energy approach.

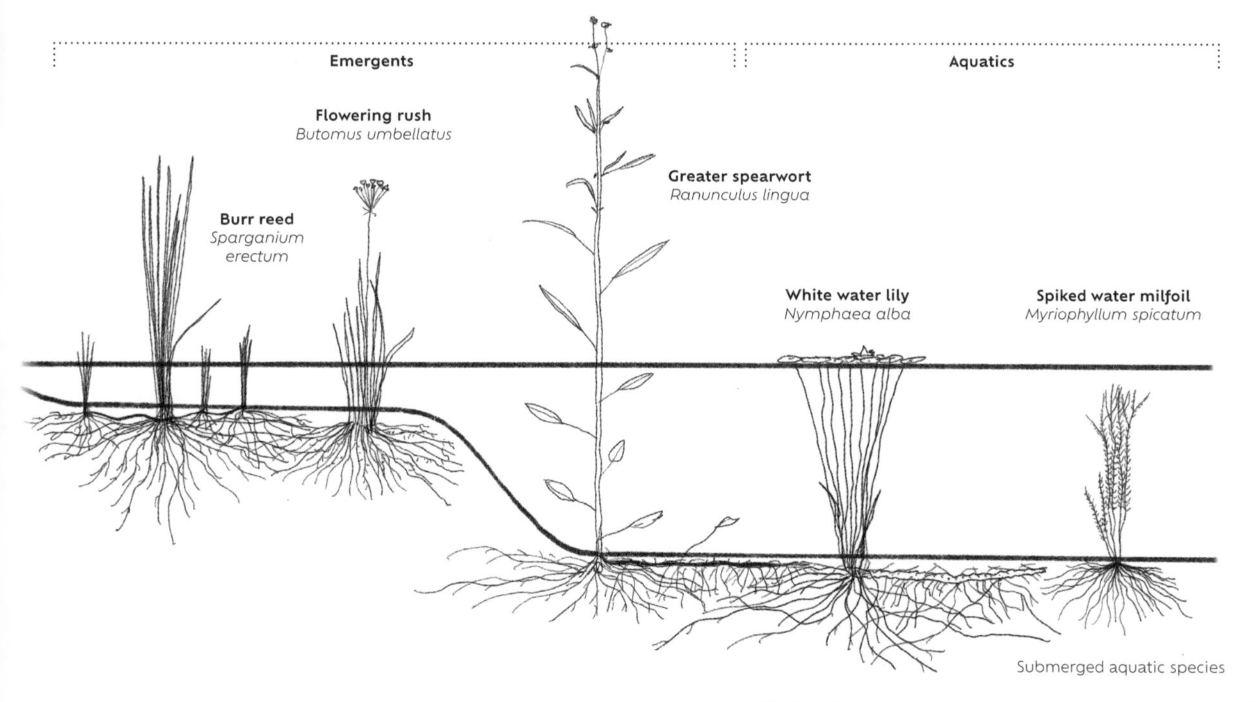

HARVESTING RAINWATER FOR YOUR LAND

> *Life is water dancing to the tune of solids. Without that dance, there could be no life.*
> Gerald H. Pollack

Water is a vital resource. In some UK areas water neutrality planning laws already insist that new house plans demonstrate that additional strain will not be put on groundwater reserves. In drought, stored water is a godsend, but any rainwater collected any time will benefit both the national infrastructure and your wallet. With no added chlorine or chloramine, your underground microbial gardeners will also thank you since, like most dogs, they would rather drink that than treated tap-water, which harms them. On a larger scale, ponds and lakes are great ecosystems: for cooling the atmosphere and creating water source heat; and for wildlife, education, wild swimming and messing about in boats. Support local wetlands and wet woodlands if you can; help bring back beavers, water voles, wild birds and other ecosystem engineers where space and licensing allow.

How to harvest enough rainwater for a productive garden

1. Look up your average annual average rainfall and calculate how much rainwater you might collect by measuring your roof surfaces. For example, if you have 100m^2 (1,075 sq ft) of roof and average rainfall of 50ml (1¾ fl oz) of rain a month, you could collect 5,000 litres (1,100 gallons) a month; allow for 15 per cent evaporation, so 4,250 litres (935 gallons). Area of roof × rainfall = volume of water you might collect.

2. Choose the best ways to collect water for your land:
- roofs – house, potting shed, bike shed, greenhouse or clean hardstanding;
- gutters and chain gutters;
- ponds and lakes;
- consider reusing grey water (see Appendix 2: Grey Water Recycling, page 239).

3. Consider storage options:
- water butts or troughs – an average water butt is about 200 litres (43 gallons);
- underground rainwater harvesting tanks – a rainwater harvesting tank might be 5,000 litres (1,100 gallons);
- pond, lake or bog area – can be much bigger – decide how much to allocate for watering the garden and how much to keep for wildlife.

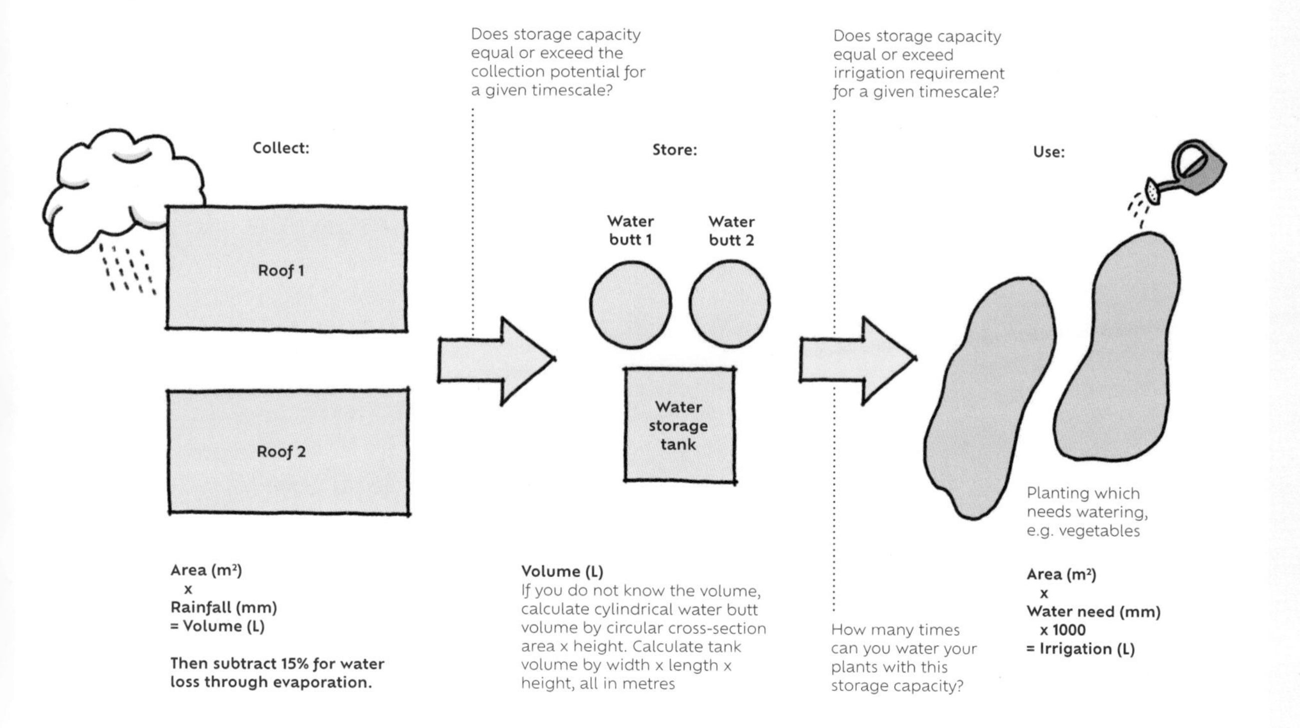

Use this simple calculation to know how much rain water you can collect and store as a budget to plan your planting.

Opposite, clockwise from top: gravel or drought-tolerant planting vegetables and wildflower meadows will all draw differently on your water budget.

4. Slow the flow to keep the water on your land so it can percolate back to your aquifers:

- green roofs;
- reed beds;
- swales;
- ponds;
- dew ponds;
- French drains;
- meanders in streams, swales and rivers – add leaky dams where possible.

5. Think how much water you need. The first five in the following list of typical types of planting will need watering for the first season and then should do without additional water after establishment. The second five will need watering. A vegetable garden typically needs 0.12mm (1⁄200in) per square metre (11 sq ft) per month in summer months, depending of course on what you are growing. For a 5m² (54 sq ft) border say, this would mean 600 litres (130 gallons) a month. A hose or sprinkler uses about 1,000 litres (220 gallons) an hour, so this equates to using a hosepipe for about 2.3 minutes a day over the month. New trees will need 80–100 litres (17–21 gallons) per month depending on size and species.

Here are some typical types of planting that you may have in your garden:

- gravel/drought tolerant;
- steppe-style planting;
- prairie;
- forest garden;
- wildflower meadow;
- vegetable garden;
- herbaceous border;
- greenhouse;
- new trees @ 8cm (3¼in) girth;
- new trees @ 35cm (14in) girth.

6. Using the rules of thumb above, draw up a water budget to ensure you save enough rain to cover summer watering needs.

7. Create healthy living soil by feeding your microbes. Your soil can become like a sponge, holding on to any water that arrives and allowing it to percolate slowly back to your aquifers.

ECOSYSTEMS: THE NETWORK OF LIFE

Economy without ecology means managing the human nature relationship without knowing the delicate balance between humankind and the natural world.
Satish Kumar

The word ecosystem comes from the Greek *oikos*, meaning household. A system is an interconnecting network that works as a whole. Jainian monk and environmentalist Satish Kumar once explained to an audience of economists (from the Greek *oikonomia*, 'household management') that their real job was the management of the whole household of the natural world, not only money – otherwise they would just be money-managers. If we do not look after the biological eco-home then there will be no money, nor people to manage it for.

Alexander von Humbolt, Prussian botanist, naturalist and explorer in the eighteenth century, was one of the first to study the way creatures live in symbiosis with their environment. Having travelled the world, collected and documented around 60,000 different species, his greatest, ground-breaking discovery was the concept of the 'unity of nature' – the idea that everything on the planet is interconnected in 'one great whole'. More recently James Lovelock's 'Gaia theory' described the universe as a single, self-regulating living entity, wherein each action has a ripple effect through the whole.

Within our planetary whole we interact intimately with many smaller wholes. Our bodies are a whole ecosystem, and within that our gut is another. Outside our bodies we have the ecosystems of our homes, gardens, towns, landscape areas and continents. As humans we define boundaries to make these areas easier to understand and to manage. The boundaries are

Above: Plants communicate with insects. Butterflies can 'taste' plants with their feet.

Opposite: Plan for maximum microhabitats, hotspots, stepping stones and corridors to allow wildlife to thrive.

blurred though, and some creatures will be happy in more than one ecosystem.

The five volumes of *British Plant Communities* (1991–2000) edited by John Rodwell are a testament to how difficult it is to generalise an ecosystem and that is partly because each is always evolving. As soil moves through succession from bare rock to ancient woodland over hundreds of years (see Soil succession, pages 48–9), so a man-made garden pond will naturally silt up, and trees will seed into it, fall and rot, creating earth for other plants to succeed. Until an event like drought, flood, fire or human intervention, the pond and its inhabitants will adapt as it becomes wetland and sometimes dry land.

Recent studies in epigenetics show how creatively organisms adapt to different environments over time. Humans who survive famine pass on a gene expression of hoarding available calories to the next generation; weeds become resistant to glyphosate; hermit crabs live in plastic bottle tops. These studies show how changes around us can impact our gene expression without changing our DNA. As our ecosystems evolve, we adapt, increasing the chances of our survival.
Both traits and traumas developed as responses are passed down to future generations. By understanding responses to environmental cues, and the relationships between microbes, plants and insects, we can begin to create adaptive habitats to increase survival of species as our biosphere changes.

The classic description of an ecosystem divides areas into biotic components like plants, animals and microorganisms, and abiotic ones, which are the surrounding physical environments, including the water, air, habitat and climate. A low-altitude seaside ecosystem is different to a mountainside, for example, so when we create planting plans it helps to know where a plant would naturally occur. When walking up a hillside, the flora changes in increments of as little as 10m (33ft). Exposure, altitude, wind, water availability – both in the soil and in the air as moisture – are all key factors. Sometimes we can find a foreign environment that perfectly mimics the home conditions of a

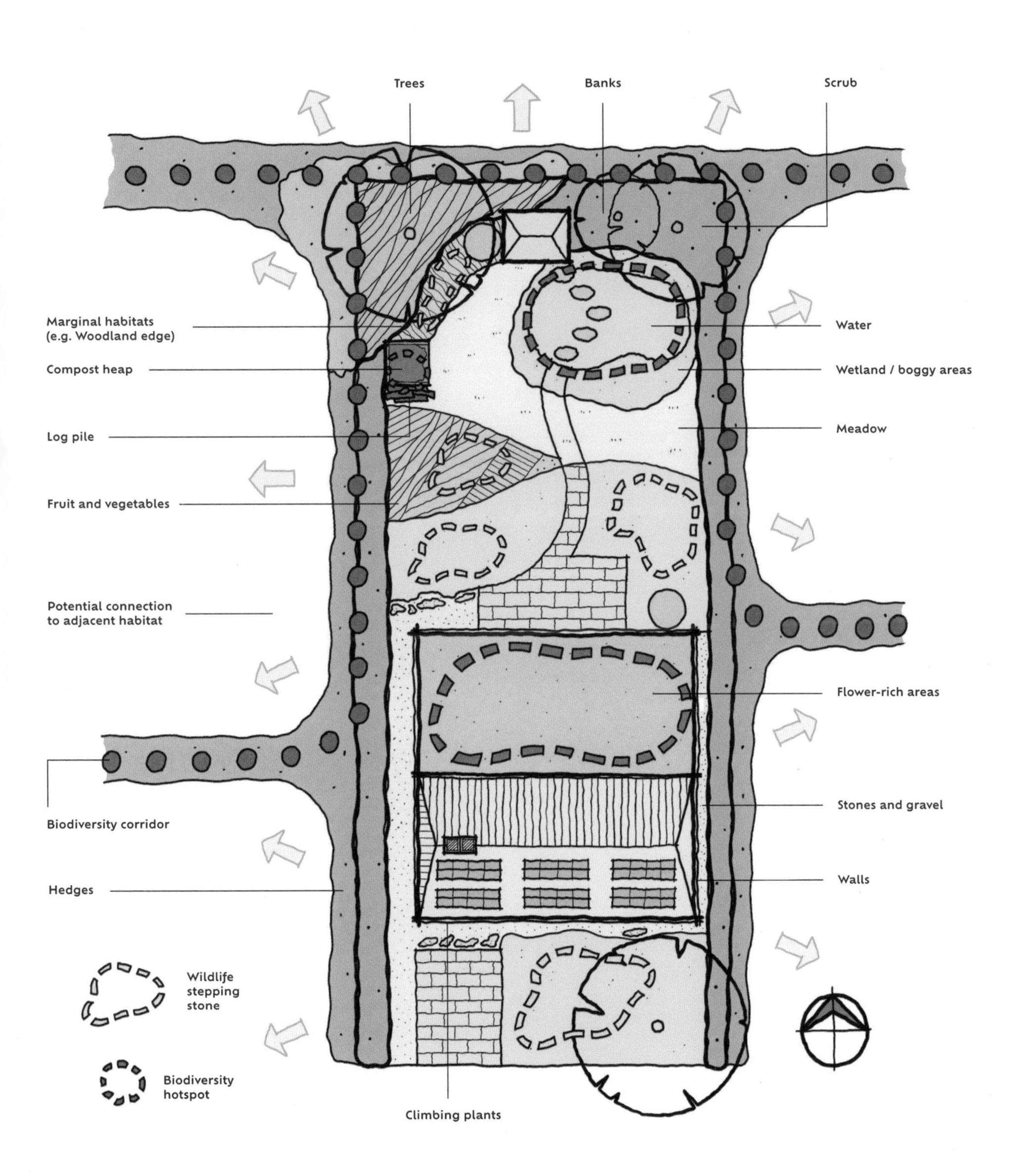
Trees
Banks
Scrub
Marginal habitats
(e.g. Woodland edge)
Water
Compost heap
Wetland / boggy areas
Log pile
Meadow
Fruit and vegetables
Potential connection
to adjacent habitat
Flower-rich areas
Biodiversity corridor
Stones and gravel
Walls
Hedges
Wildlife
stepping
stone
Biodiversity
hotspot
Climbing plants

plant: on the Tregothnan Estate in Cornwall, for example, the land is well drained, near sea level, faces south and has a sea mist that creates the perfect conditions for growing tea (*Camellia sinensis*), which usually prefers the humid hillsides of Sri Lanka or China. Similarly the great cook and gardener Mark Diacono looked at a toy globe with his daughter and realised that he lives on the same latitude and altitude as Bellingham, near Seattle, Washington, and so could grow their same northern-acclimatised pecans in Somerset.

Within each ecosystem there is a web of dependencies and energy exchange. Starting from the bottom up, down in the soil in deep darkness, without light or need for eyes, the microbes have extraordinary sensory systems to determine food from friend or foe and to communicate through a system of 'infochemicals' that we are just beginning to understand. The microbes interact with the plants, which in turn work with sunlight through photosynthesis to produce sugary exudates; these they feed to the microbes in exchange for nutrients from the soil.

By understanding relationships between microbes, plants and insects, we can create adaptive habitats to increase survival of species as our biosphere changes.

Above ground, plants communicate with the insects they feed and rely on for pollination, through bio-resonance or vibrational signals, chemical signalling and colour signalling. To attract a pollinator, a plant might vibrate at a minute resonance far beyond human perception but loud and clear to a bee. When a plant like the horned pansy (*Viola cornuta*) or tea rose 'Mutabilis' (*Rosa* × *odorata* 'Mutabilis') has been pollinated, it changes flower colour to signal to pollinators to go elsewhere. Some plants attract insects by mimicking their mates: for example, the bee orchid (*Ophrys apifera*) emits a pheromone to lure males in. Insects, like a butterfly, can 'taste' a plant through their feet or, like a bedbug, detect the heat of a human body through their antennae. Plants also use chemicals such as indole to ward off being eaten. This is a signalling molecule, which we recognise as the smell of faeces. It attracts both natural predators and parasitoids of insect pests and so can regulate the plant's defence systems, repelling pests.

Plant breeding has altered a plant roots' ability or incentive to communicate. Some cultivars, such as newer wheat varieties, no longer signal to beneficial protists for nutrient cycling or bacterial control, and they no longer initiate symbioses with mycorrhizae.

The use of insecticides, fungicides and herbicides impacts photosynthesis as well as root exudates. Neonicotinoids have been found to alter the genes responsible for cell wall structure, detoxification and switching on enzymes and phytohormones involved in defence.

Gardeners can help reverse some of these gene expressions by applying living compost, vermicast from a wormery, seaweed or plant extracts like humates and compost teas (see page 63), to put back microbial health and support plant elicitors, the compounds that activate chemical defences.

Regenerative gardening involves a mix of biology, chemistry, physics, geology, climatology and social sciences, with added layers from systems thinking and epigenetics. A syntropic approach involves sharing knowledge across disciplines, so I always work with ecologists to establish what and who is living where and how best to minimise harm and add benefit to their biota. On a nature recovery scale, ecologists like Derek Gow have pioneered the reintroduction of beavers and storks to the UK and the recovery of smaller species like glow-worms and harvest mice. He and others like Andy Phillips are doing crucial research in developing effective strategies to regenerate our ecosystems. Their work at Great Dixter in East Sussex, John Little's Hilldrop in Essex, and my own garden in Kent have shown the value of domestic gardens to wildlife regeneration.

It is vital to understand some of the roles and functions that make up ecosystems, but I am wary of the trend to quantify and monetise ecosystem services. Some people argue that this confers a financial value that may protect them, and yet metrics like BREAAM and Biodiversity Net Gain (BNG) can also oversimplify a complex natural order and, if we are not careful, become another set of numbers to be gamed for human gain. The syntropic systems in nature show life as a network, where species exchange goods and services for mutual gain, and one system's waste becomes another's food. This is the long-term balanced and self-regulating whole that Von Humbert and Lovelock described. How wonderful it would be if we could all participate in that fully reciprocal relationship.

UNDERSTANDING GARDEN ECOSYSTEMS

> *The breeze at dawn has secrets to tell you.*
> *Don't go back to sleep.*
> *Rumi*

When making a garden, bed or border we plan structure, shape and year-round interest in colour, scent, texture and interaction with light and shadow. Spreadsheets and layout plans are useful to ensure soil cover, synergy between layers, and services like nitrogen-fixing and nutrient cycling. Once we have mastered that complex art, there is a whole next layer of vibrant life and interest to integrate and help curate.

Understanding who else shares our garden and what they need to thrive brings us a step further in tune with the nature we are part of. Like us they need sun, shade, water, food and shelter.

Buildings, planting and landform provide shade. Ponds, puddles and shallow dishes of rainwater are lifesavers for many creatures, especially in hot weather. Before we talk about forage, let's look at shelter. We provide food for birds and flowers for pollinators, but where do they live?

Birds are either residential or migratory, travelling south to feed in winter and returning to raise their chicks. All need nests in spring, to raise one or two broods of young, after which they roost in trees, hedgerows or buildings for the rest of the year. The nest will vary from a precariously messy pile of sticks on a branch for a wood pigeon, to the neat, cup-shaped circle of mud, sticks and moss of a blackbird inside a hedgerow. Songbirds like blackbirds, thrushes and bullfinches all like blackthorn (*Prunus spinosa*) and hawthorn (*Crataegus*) hedges, and a tall tree nearby to sing from. The thicker the hedge the safer they are from the cats, rats, foxes and larger birds who feed on them.

Once established they will fill early mornings with song and be head of the integrated pest management (IPM) team. The wren is one of the UK's most common, smallest and loudest birds, which nests in cracks in walls and crevices in trees and will eat spiders, flies, millipedes, ticks and lice, aphids, beetles and ants, caterpillars and other insect larvae, small seeds and will even fish for tadpoles. One male wren in the milder south of the UK will build up to five nests for different females, who

Like us our wildlife need sun, shade, water, food and shelter.

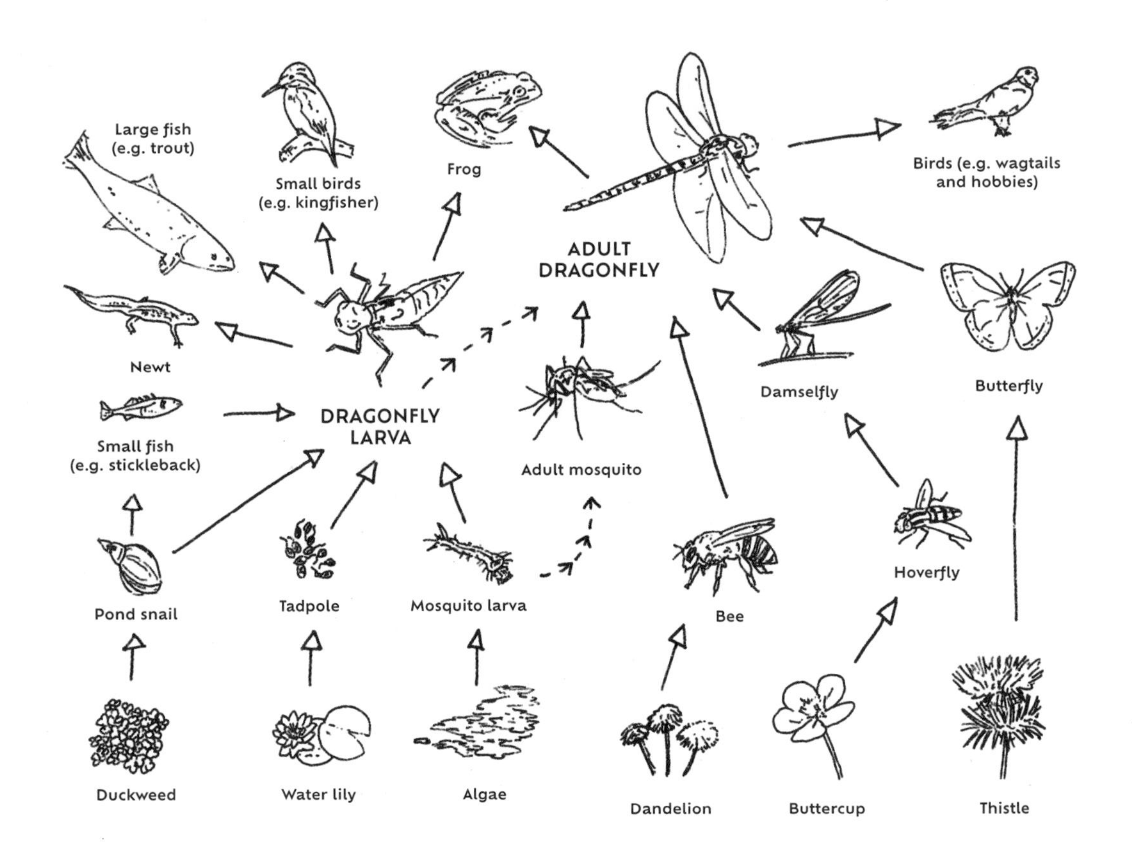

Above: A garden ecosystem will have several different microhabitats, which interconnect and overlap, as shown by this dragonfly within its food web.

Opposite: *Lasioglossum*, one of our smallest UK native bees, feeding on catmint flowers.

inspect and choose the best, which usually means best hidden. In a cold winter families will roost together to keep warm, with over 60 sometimes found together. Nest boxes are helpful in towns, but even warmer and safer are crevices in garden walls and structures, so keep these unfilled, and protect old hedgerows, plan new ones, and allow old trees and dead wood to stand.

A garden ecosystem will have several different microhabitats, which interconnect and overlap. Try to understand your garden from the point of view of the animals, plants and insects. Let's take the dragonfly as an example within the food web. Adult dragonflies and damselflies eat flying insects like mosquitoes and in turn have many predators: frogs, spiders, fish (when they are laying their eggs), ants (when newly emerged) and birds. The larvae are eaten mainly by fish, water beetles and some birds. Larger larvae will also eat smaller larvae. To create their habitat we need a pond with floating plants like waterlilies (*Nymphaea alba*) and water soldiers (*Stratiotes aloides*) to provide shelter and shade for the larvae, and mating sites and tall-stemmed plants like flowering rush (*Butomus umbellatus*) and pickerel weed (*Pontederia cordata*) from the USA for the larvae to climb up and

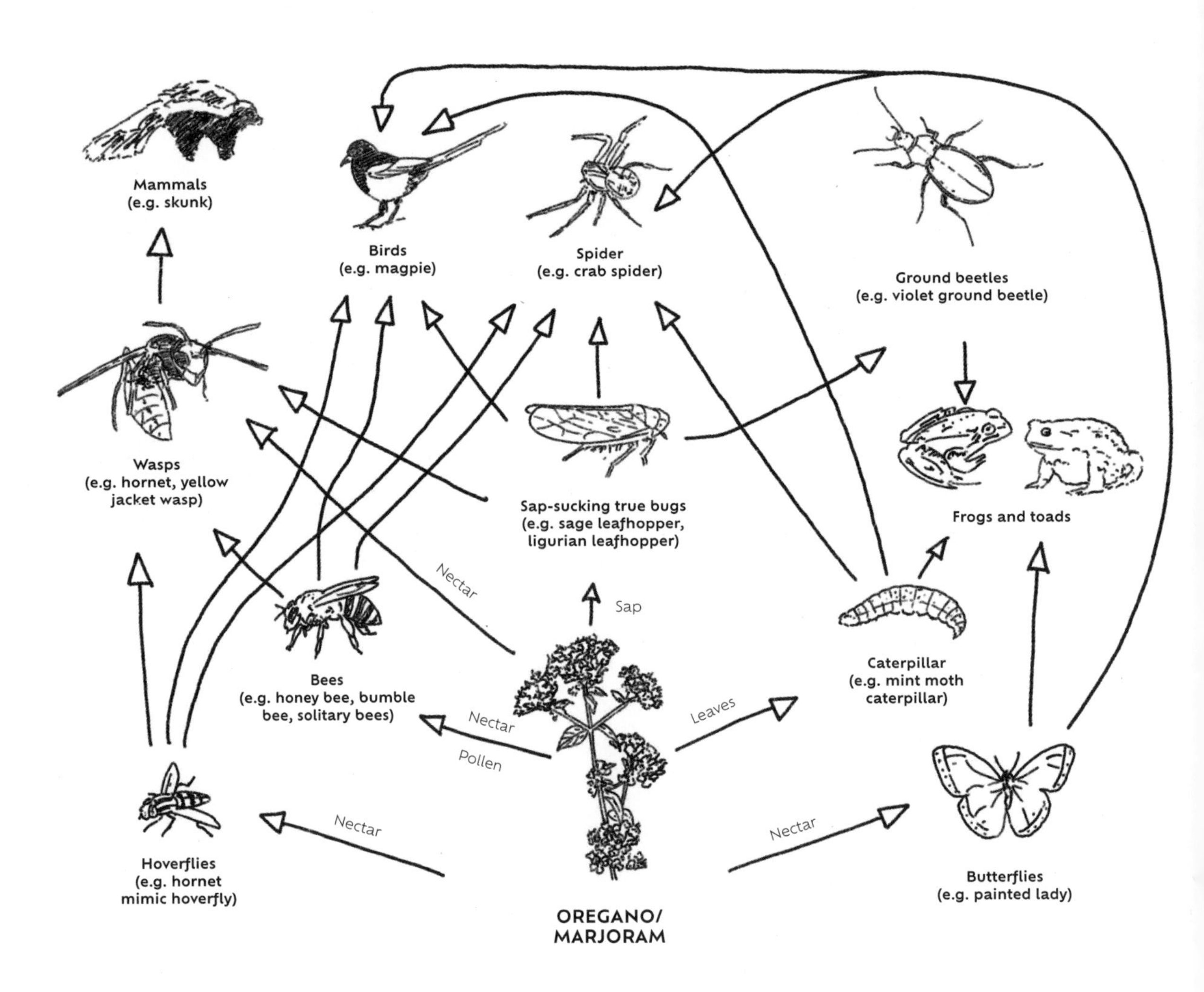

Mammals
(e.g. skunk)
Birds
(e.g. magpie)
Spider
(e.g. crab spider)
Ground beetles
(e.g. violet ground beetle)
Wasps
(e.g. hornet, yellow
jacket wasp)
Sap-sucking true bugs
(e.g. sage leafhopper,
ligurian leafhopper)
Frogs and toads
Nectar
Sap
Bees
(e.g. honey bee, bumble
bee, solitary bees)
Caterpillar
(e.g. mint moth
caterpillar)
Leaves
Nectar
Pollen
Nectar
Nectar
Hoverflies
(e.g. hornet
mimic hoverfly)
Butterflies
(e.g. painted lady)
OREGANO/
MARJORAM

Opposite: The simple oregano (*Origanum vulgare*) is the central source of life in this illustrative food web.

Above: At the centre of all of food webs are the plants, which convert chemicals – from the air and soil and minerals from soil microbes plus the sun's energy through photosynthesis – to energy for all of us.

pupate into emerging adults. If the pond has frogbit (*Hydrocharis morsus-ranae*) this will feed pond snails, which also relish dead plants and are food for the dragonfly larvae and for small fish. In turn the small fish may eat the dragonfly larvae and also mosquito larvae (which will live on algae) and tadpoles (which hide under leaves like waterlilies).

At the centre of all of food webs are the plants, which convert chemicals – from the air and soil and minerals from soil microbes plus the sun's energy through photosynthesis – to energy for all of us. In addition to using some of the energy for growth and some as plant exudates to feed soil microbes, flowering plants also develop flowers to reproduce. The male stamens produce pollen – tiny yellow granules of male gametes that need to land on female pistils of another plant (unless self-fertile) to germinate and create a pollen tube down into the female ovule. There the pollen capsule opens to deposit the sperm so it can fertilise the female gametophyte. To disperse the pollen between flowers, around 12 per cent of plants need wind, while the remaining 78 per cent rely on insects.

Some plants offer sweet nectar to insects and, while they drink, the plants deposit pollen on the insect's head or hairy bodies, like those of mason bees and leafcutter bees. The pollen will then be brushed off on another plant, and hopefully on to the pistil. Many insects like bees also collect pollen in sacs on their legs or abdomen as a protein to feed their young. If they collect too much, not enough goes to pollinate plants, and if they mix it with nectar and saliva they make it non-viable for reproduction, which is one reason why a diversity of pollinating insects is important.

The Western honeybee is the main bee pollinator worldwide and meets about 34 per cent of current crop pollination requirements in the UK. Of our 267 species of bee in the UK, this is the only domesticated bee and, where land is farmed intensively, it is vulnerable to diseases and so can disrupt natural pollinator networks.

According to some studies wild bees without pollen sacs can be up to 40 per cent more effective as pollinators than honeybees and bumblebees, since they disperse pollen as they fly. Generally smaller than honeybees or bumblebees, these include specialists like mining bees (which make nests in sandy ground), leafcutter bees (which cut sections of leaf to build nests in dead plant stems) and mason bees (which nest in crumbling bricks and walls). Unlike the hive-dwelling honeybee, once these bees have laid an egg with its provision of pollen in a deep hole and sealed it up, many take no further part in their offsprings' lives; the offspring overwinter and emerge in spring to live for 4–6 months and complete the same cycle. This makes them quite vulnerable in winter, and several cuckoo bees have ingenious ways to remove eggs and replace them with their own or just add their own; the cuckoo bee offspring will eat the food and the host egg or larvae, or just eat the larvae. Like all insects they are also vulnerable to humans tidying up and destroying their nest sites.

Hoverflies are very effective pollinators. They look like wasps but have large, fly-like, composite eyes, which meet in the middle in some males, and possess no sting. They feed on nectar and pollen, and their larvae feed on aphids, thrips and leafhoppers or decaying vegetation, so are a useful part of an integrated pest management system for growers.

Ants are pollinators too, and can aerate soil even more than earthworms as they form nests. Intelligent group-working creatures, they have synergistic relationships with some gall-forming wasps, which lay their eggs in galls on plants like oaks (*Quercus*) and attract ants with the nectar secreted by the galls, in return for which the ants protect the galls from predators. Ants also milk aphids' sweet honeydew by stroking their abdomens, keeping them in herds subdued with chemicals excreted from their feet while protecting them in return from predators like ladybird larvae.

Butterflies, moths, flies and beetles also have a big role to play in ecosystems. Butterflies and moths are pollinators by day and night, respectively, and can pollinate flowers with longer corollas as they have long tongues (probosces). They are food sources for birds, hornets and spiders as adults and for toads, birds, snakes and predatory wasps as caterpillars. Voracious eaters of leaf material in the larval caterpillar phase, they can

Provide homes for ground dwelling invertebrates, as these are the vital clean up crew.

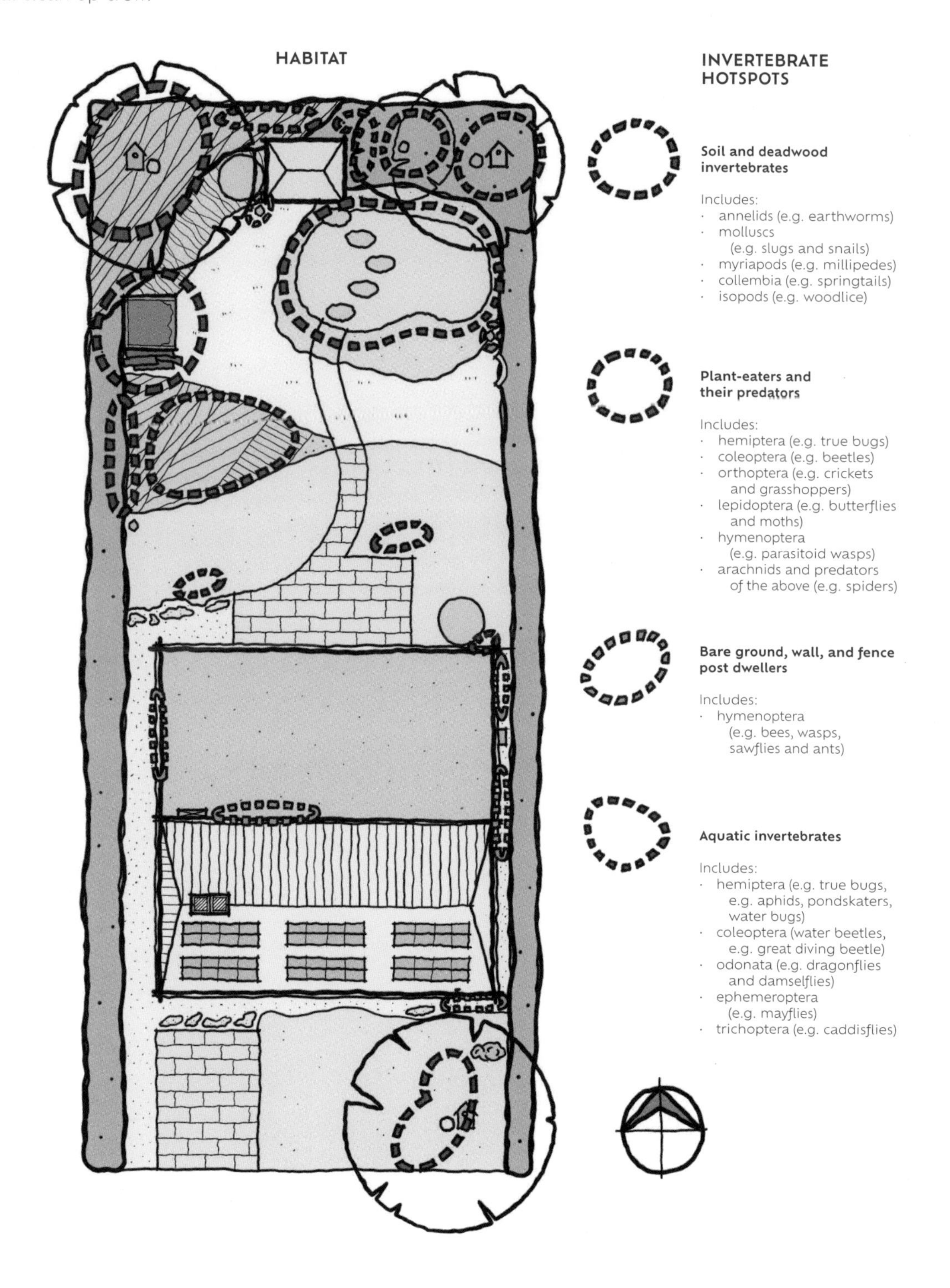

Opposite: clockwise from top left: pollinators like this holly blue butterfly, hoverfly and tiny mint moth are a vital part of the garden ecosystem. You can tempt caterpillars away from your prize brassicas by planting nasturtiums.

eat up to 70,000 times their body weight as they change skins four or so times before they pupate into butterflies. For a painted lady butterfly this takes only four weeks, while a goat moth caterpillar may remain inside a tree trunk for up to five years.

Plants adopt several strategies to work with insects. For example, they are careful not to give out so much nectar in one go that an insect is sated and just goes home instead of on to the next flower. Some flowers will make an insect walk around the whole flower gathering small bits of nectar as they go from each nectary at the base of the flower (and spreading pollen on each anther). Pollinators can take short cuts too: some will pierce the nectary of a tubular flower like sage (*Salvia*) rather than entering to pollinate.

Beetles and flies are part of the vital clean-up crew (detritivores) feeding on dead plants and animals and faeces and recycling nutrients. The importance of dung beetles cannot be overstressed in the countryside. They are key ecosystem engineers, and in regenerative farming systems they prevent disease in animals and allow fields to be productively grazed and nutrients recycled. Dog and cat owners can help bring these back by avoiding insect-killing wormers like Ivermectin, using them only if animals show signs of worms, picking up poo and disposing of it in landfill if their pets are treated. Spot-on chemical flea treatments, which contain devastating neonicotinoids, can also be replaced with natural products like 'Billy No Mates' and used prophylactically.

With approximately 1.4 billion insects for every person and the total weight of all insects at about seventy times more than all the people, we are learning more everyday about these vital creatures that support and delight our lives on Earth.

ECOSYSTEM ENGINEERS

> *The more complex the network is, the more complex its pattern of interconnections, the more resilient it will be.*
> *Fritjof Capra*

Ecosystems are dynamic. Every new plant that arrives will change the community. Some may fit right in and balance will be restored, while others will be over-dominant and take space from the existing palette. Seeds are brought in by the wind, birds and ants – and of course by humans. Many of our plant and animal introductions have been helpful, like the European honeybee to North America and European dung beetles to Africa, which were vital to clean up after the imported cattle. Others have been useful but annoying to gardeners, like ground elder (*Aegopodium*), brought to the UK by the Romans to cure gout, and bamboo (*Bambusa*) from China, a fast-growing biomass also difficult to contain.

According to a paper published in the ecology journal *Oikos* in 1994, ecosystem engineers are 'organisms that modulate the availability of resources for other species, either directly or indirectly, by altering the physical state of biotic and abiotic materials within their ecosystem'.

Beavers, water buffalo and bison are all excellent large ecosystem engineers, whose role is to make different shapes in the landscape, creating habitat and making nutrients available. Beavers do this through building wetlands. They are herbivores that mostly eat willow (*Salix*) shoots, and bring down trees so that they can reach the tasty tips. They also turn branches and mud into dams to have safe wetlands to breed. The increase in wildlife in the habitats they engineer is phenomenal, as well as holding water in the landscape, which stores carbon and becomes a buffer for times of drought. Water buffalo dig scrapes and wallows and expose soil for insects and birds as well as holding some water like the beavers. Bison and wild boar also help to generate through destruction and defecation: by knocking down trees and rubbing up against bark they actively regenerate ecosystems, while rootling and pooping in the earth enables insects and fungi to flourish and with them all the plants and animals that rely on them in their food cycles.

Humans are, of course, the busiest of engineers, and we have the advantage of being able to invent, plan, collaborate and

Wild boar rootling and pooping in the earth enables insects and fungi to flourish, as well as all the plants and animals that rely on them.

Above: Old gate posts, hollow trees and standing deadwood are perfect homes for bees.

Opposite above: Leave acorns on fence posts for jays to take and plant.

Below: The dung beetle deserves special recognition as the king of the detritivores.

marshal resources on a vast scale. In recent times we have modulated the resources in our short-term favour – often to the detriment of our cohabitors – but it is possible to become a beneficial engineer on a garden and a landscape scale.

In the garden one of the smallest engineers that we can emulate is the ant, which makes lumps and bumps and aerates the soil. If you are lucky enough to have ants take a look at their handiwork, and if not, you can build some mounds of earth or sand and reclaimed aggregate as mini-habitats that have different slopes and aspects for tiny creatures and plants to colonise.

Cater for bees, who pollinate our plants and so provide food for us all, by leaving hollow trees and standing deadwood, old gate posts and cracks in mortar between bricks, and planting year-round pollen and nectar for them. Slow the water like the beavers, creating leaky dams in streams and rills and allowing floodplains and dew ponds to form naturally, or create shallow scrapes and wallows like the water buffalo. Be like the jay and reseed woodlands and hedgerows, or leave out acorns on fence posts for real jays to take and plant with a lunch box of poop to get them started.

The dung beetle or scarab is revered in Egyptian mythology as the god of rebirth Khepri, and deserves special status as the king of the detritivores, without whom we would be wallowing in undecomposed waste and sickness. The dung beetle's vital role is taking the nutrients down into the soil to become natural fertiliser. Provide habitat for these creatures by allowing areas of slow decomposition, where wood is left to rot gradually, and by bringing animals in and letting their waste be trampled into the soil. Help dung beetles by using natural wormers for dogs and cats. Be a human dung beetle! Make fast compost (see Making compost, extracts and tea, page 56) to break nutrients down enough for that other great engineer – the earthworm – and our smaller underground livestock to turn them into humus.

On a larger scale, design in Hügelkultur mounds (no-dig raised beds), fill steel gabions with mixed aggregate and bricks to cater for invertebrates and lizards, install owl boxes and bat boxes, leave cracks for swifts, and install swift boxes and house martin bowls on new builds. Build sand boxes for bees and parasitic wasps and beetle banks on south-facing slopes. Make hedgehog houses and water vole and toad tunnels, and leave gaps under fences and make our hedges contiguous to enable safe passage. Provide thorny barriers of hedges with food growing in them to feed and exclude deer from special plants you don't want to share, and eat wild venison if you prefer to have fewer deer, to avoid wasting their noble lives. Pile up recycled rubble in a warm corner and cut lawns at different lengths to maximise edges. The more of this edge habitat you can establish, the more diversity you will encourage – and with diversity comes resilience.

Matt Somerville creates homes out of logs for wild honey bees to thrive.

CREATING A HAVEN: A PRACTICAL GUIDE TO PLANNING AN ECOSYSTEM GARDEN

> *There can be no purpose more enspiriting than to begin the age of restoration, reweaving the wondrous diversity of life that still surrounds us.*
> *E.O. Wilson*

You may have been taught 'right plant right place' in terms of sun, shade and soil type. Once we also begin working with where plants choose to grow, and which other plants they associate with, we open up a whole new dimension to creating a garden that functions and feels deeply comfortable, as a near archetype that we recognise at an atavistic level. The next layer then is to cater for the many animals that associate with different plants, and set the stage for a garden full of abundance and life.

Observe the insects and birds that visit areas of planting to build up a picture of the nectar and forage hotspots in your garden at any given time. They will move around the garden during the day and over the months as areas become sunnier and warmer, and plants blossom and fade. For each season plan a spread of plant shapes that are most attractive to the foragers currently in search of food. Include generalists and specialists, varying the plant shapes through the garden rather than providing a homogenous covering.

Plan for about 60 per cent generalists – magnets for lots of different insects. These are the 'party plants', like marjoram (*Origanum majorana*), knapweed (*Centaurea nigra*) and many of the umbellifers and *Asteraceae*, where nectar is on tap and everyone is invited. Often white or yellow, the umbellifers may have an evolutionary advantage with light wavelengths that most insects can see. The *Asteraceae*'s appeal is in its inflorescence: being made of lots of little flowers with short corollas, these give easy access to the nectaries. Both umbellifers and *Asteraceae* will be covered in bees, butterflies, hoverflies, flies, sawflies, beetles and butterflies throughout the flowering season.

Although having plenty of generalists means you will attract many pollinators, these may need many visits to ensure cross-pollination. Of all insects, bumblebees, followed by solitary bees, are the most prolific pollinators.

Your planting should also ensure forage for specialists, including night-scented flowers for moths. Specialist flowers are particularly attractive to a few bee species: for example, bellflower (*Campanula*) offers food to bellflower blunthorn bees and small scissor bees; yellow loosestrife (*Lysimachia vulgaris*) supports yellow loosestrife bees, which collect the pollen and also the oils to waterproof their nest cells.

Plants with tubular flowers like winter-flowering honeysuckle (*Lonicera fragrantissima*) or later-flowering trumpet vine (*Campsis radicans*) are a draw to long-tongued bees and moths, which can reach down into the corolla, as well as to some 'robber' bees, which will pierce the corolla at the top to reach the nectar. Legume pollen is a rich source of protein for pollinator larvae, so plan a mix of *Fabiaceae* to suit different tongue lengths of foragers and so benefit insects and nearby farmers. Summer-flowering crimson clover (*Trifolium incarnatum*) will support long-, medium- and short-tongued bumblebees as well as some hoverflies, while red clover (*T. pratense*) feeds long- and medium-tongued hoverflies.

A spread of plant shapes for the different foragers in each season. Include generalists and specialists, varying plant shapes through the garden.

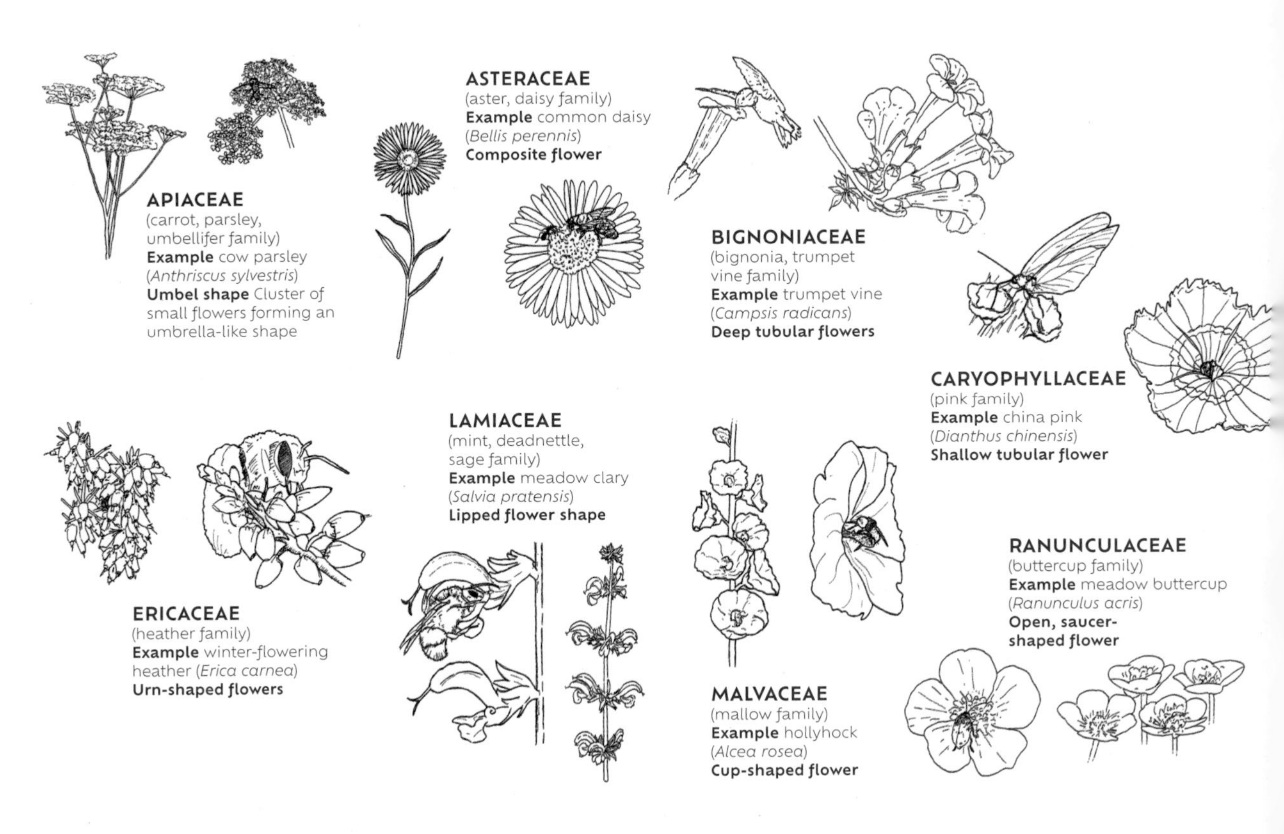

Pollinator hotspots in summer:
Top plant families for forage.

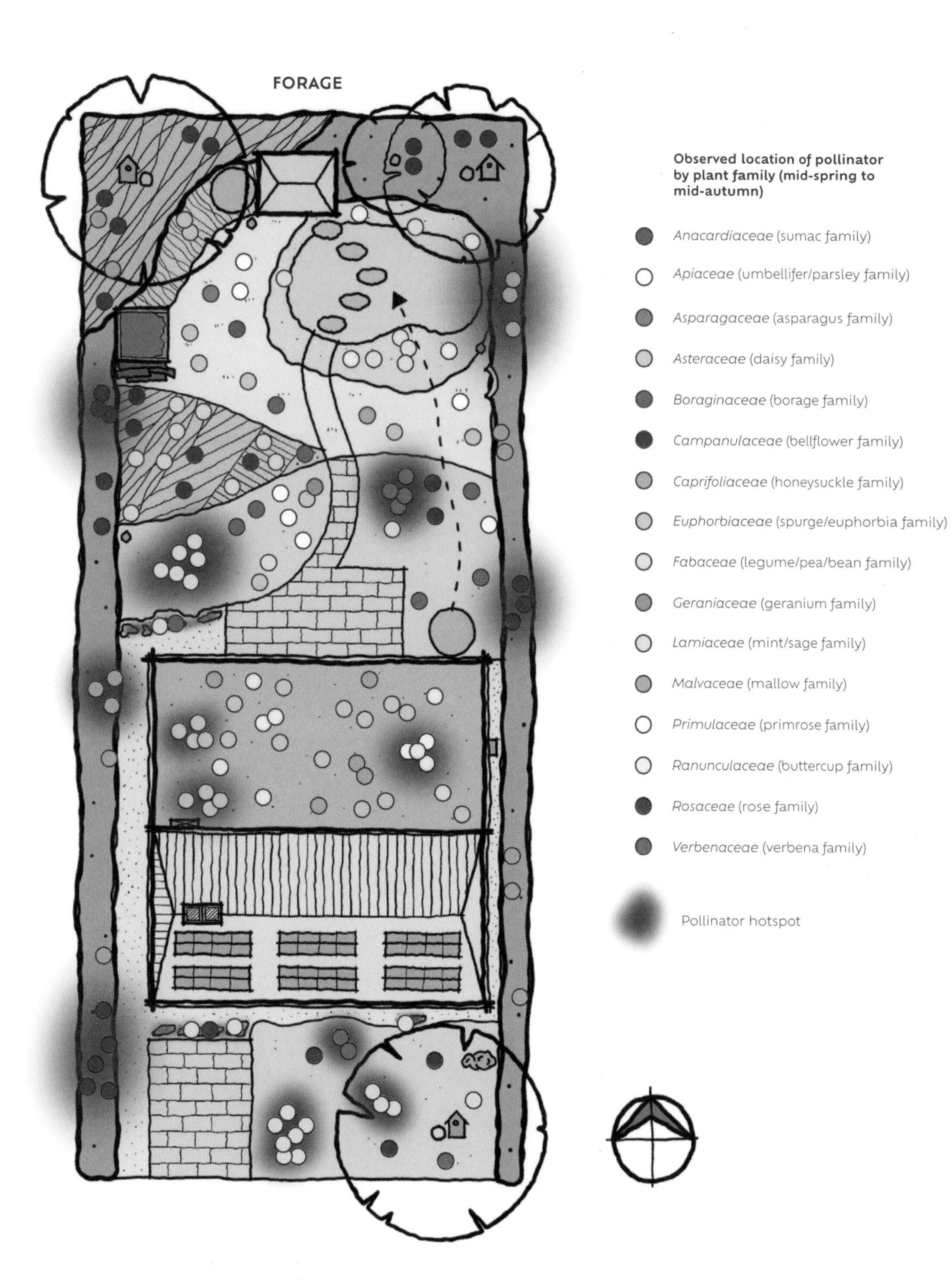

Opposite: Build a variety of bee nesting sites through the garden, like this simple log pile (see page 120).

Trees, shrubs and climbers are also important. Listen to the buzz of mature ivy (*Hedera*) in flower in late summer – a nirvana to ivy bees, while early-flowering willows (*Salix caprea* and *S. alba*) are mobbed by hungry bees emerging from hibernation in spring.

The purpose of pollination is to set fruit, seed and reproduce. Flowering at different times of year helps, both by allowing enough time to ripen fruit and disperse seed, and by flowering when the right pollinators are available. The bee orchid tempts the male bee to land on its female-bee-mimicking pad by emitting pheromones, but gives no nectar. This strategy wouldn't work for long, but the bee orchid flowers when the young males are just starting to fly, so benefits from their naivety. Not all plants need pollination, however: some beans and peas are self-fertile, and grasses like wheat and oats are wind-pollinated. Some fruits and vegetables are self-pollinated but benefit from additional 'buzz' pollination to produce edible fruits. Tomatoes and aubergines hold their pollen in hollow anthers that are 'shaken' by bees' high-frequency wing vibrations to rain on to the flowers below. It is not understood why bees buzz exactly when they do, but we do know that bumblebees and solitary bees do this, so position bee-attracting companion plants like garden catmint (*Nepeta × faassenii*) by an open greenhouse.

Build a variety of bee nesting sites through the garden to suit these visitors. These can range from the hollow stems in cow parsley (*Anthriscus sylvestris*) to dead bark for beetles. Erect a vertical compost bay to allow overwintering insects to stay in hollow stems and emerge in spring rather than succumb to rot in a compost heat or be chipped or burnt on a bonfire. Leafcutter bees like to make nests in old fences, plant stems and dead wood and line their cells with pieces of leaf from roses and wisteria glued together with saliva.

Bee hotels are useful in gardens without these homes, in which case tubes should vary in diameter and be 20cm (8in) deep to allow for the female larvae to be encased deep inside. Shorter lengths may result in only males.

Detritivores are key to any circular system and your compost heap and woodpile will provide good habitat, as will leaving wood on the ground and dead bark for beetles. Make sure to include water. As the dragonfly food web example showed (see Understanding garden ecosystems, page 98), water is vital to an ecosystem as an algal food source as well as for the amphibians

A thriving garden ecosystem will also include a variety of co-existing mammals, so find your own balance between tidiness and tolerance.

it will support, whether the water is in a lake, a rill or a bathtub pond. Don't rush to put fish in a small pond, though; they can be overzealous predators on both plants and frog and newt tadpoles and can raise nutrient levels beyond those that can be recycled. (See A clay-puddled pond, page 83, for planting ideas.)

A thriving garden ecosystem will also include mammals, from field mice and shrews to increasingly rare harvest mice and hedgehogs, and to badgers and foxes. Find your own balance between tidiness and tolerance. Cover and connectivity are the key needs, so establish areas of scrub in larger spaces or woodpiles in smaller ones for habitat, plus longer grass and dead hedges contiguous with live hedgerows, for corridors between forage hotspots and homes.

I am often asked about the 'problem' of mice, and my answer is owls. Mice eat seeds, fruit and insect larvae as well as human leftovers and are in turn a food source for the tawny owl, barn owl, buzzard, fox and badger. In the greenhouse a mouse-proof winter seed tray is a great boost for human–mouse relations. If the mice are too hungry in the vegetable garden, peppermint oil helps to send them elsewhere.

Spiders are a vital part of the ecosystem, and I refer any fearful readers to the children's book *Charlotte's Web*. These extraordinary creatures help reduce flies and moths in our houses, and in the UK very few will bite and none is poisonous. I was once tempted to use a sonic mouse repellent, which emits a high-pitched noise to repel mice, in my 1490s house. I didn't realise that it repelled spiders as well, until we noticed an influx of houseflies in droves. The plug-in kit was quickly recycled!

In nurturing ecosystems our aim is to increase resources, to put back into a syntropic system and help more life to thrive. On a large scale this might be reconnecting a floodplain to its river, by paying attention to the way the water flows and following the path of least resistance. This gentle morphology translates well to a garden setting, where we can observe, see what is needed and provide it. The simplest of these is adding shallow water scrapes or bowls for dry weather. See where birds and bees drink from puddles and ensure some water every 10–30m (33–100ft) – the maximum distance a small mammal will run across an open space.

Reducing hedgerow cuts to every three years is a stipulation in higher-tier countryside stewardship on farmland, and a good idea in residential gardens too. This allows songbirds greater protection from predators and allows fruits to form, feeding hungry migratory birds like redstarts as well as providing humans with forageable hips to ward off winter flu.

If there is no chance of a hedgerow, a beetle bank is a great addition to a field or meadow. A simple bund, about 40cm (16in) high and 2m (7ft) wide, will give cover for small mammals and invertebrates and has been found to shelter birds too. It's a simple structure to recreate in a vegetable garden or car park, to link with hedgerows and provide the vital connectivity for small mammals, as well as hunting grounds for larger mammals and birds of prey.

HOW TO CREATE HABITAT
SAND BOXES AND LOG PILES

At the smallest scale a simple log pile or sand box is an excellent habitat for mining bees and solitary wasps. In the tiniest of spaces a wine box will do! Simply drill a few holes to allow insects to burrow into the sides, fill with sand and position in a sunny spot. I added a couple of resilient plants – herb robert (*Geranium robertianum*) and lemon balm (*Melissa officinalis*) to the sand box – but self-seeders will also find their way in.

If you have some logs a log pile will provide a home to detritivores and mining bees as it gradually decomposes over a long time. If you can put one in the shade and one in the sun you will find different species take advantage of them.

Constructing a sand box

You will need:

- Reclaimed timbers of random lengths and widths
- Old bricks, logs and any non-toxic building items
- Horticultural sand

1. Gather even lengths and matching random widths of the reclaimed timber, for three sides of about 45cm (18in) high and a lower front for the sand box. There is no need for a base.
2. Fill the bottom with the bricks, logs and any non-toxic building items. Top up with the sand.
3. Slope the sand from front to back to allow maximum south-facing slopes for warming insects.
4. Place the sand box in a sheltered, south-facing location.

Making a log pile

You will need:

- Logs of 40–50cm (16–20in)
- Electric drill and drill bit for 8–15mm holes

1. Pile up logs in descending size to keep the pile stable.
2. Drill straight holes, 20cm (8in) deep to ensure mining bees lay female eggs. (If any hole is too short, bees will only lay the male 'guard' eggs, which they consider more expendable.)

MATERIALS: HOW TO CHOOSE MATERIALS TO WORK WITH

Buy less, choose well, make it last.
Vivienne Westwood

The materials we use to build landscapes and structures in them shape our environment. From locally sourced, natural stone to complex composites, there is a wealth of options for a variety of budgets, aesthetics and performance requirements. The wide range of products and eco-claims can make the choice overwhelming, and polarities are too often drawn, suggesting that all concrete or all greenhouse gases are 'bad' (where would we be without carbon dioxide, oxygen and H_2O?) and all timber is 'good', when in fact that depends on how it is grown, used and disposed of.

The best option is to use materials already on site, from recycled paving setts to salvaged concrete aggregate. When it is not feasible to use reclaimed, the next best alternative will depend on several factors besides aesthetics and financial cost: how far the material comes from and how it was manufactured (both in terms of materials and human impact), with some materials still relying on forced or child labour and unsafe work conditions. How easy is it to repair, deconstruct and reuse, and where will it or its components go at end of life? Will it decompose gradually into the land or emit methane in landfill, for example? Is a 'natural' product farmed in a biodiversity desert, grown as a monoculture and reliant on fertilisers and pesticides, which pollute watercourses and enter the food chain, or is it contributing to a balanced future in harmony with nature and limiting chemical usage?

The Planetary Boundaries Framework was created in 2009 by Johan Rockström at the Stockholm Resilience Centre to describe

Reclamation yards and antiques fairs are great sources for materials and furniture.

and measure the nine major boundaries within which humanity can continue to thrive for generations to come. It is constantly updated and is a useful framework for decision-making, particularly when choosing materials.

Some key issues to look at include the following.

- **Climate impact** – assessing the carbon footprint of a material over its lifetime, from raw material extraction and manufacture to installation, use and disposal. As well as carbon, consider how the colour and texture of a material will affect how much heat is absorbed or reflected. Is there any impact on land use or change of land use in producing the material?
- **Fresh-water impact** – assessing the 'embodied water' of a material. In a similar way to carbon, a material has a water footprint. Some materials use a huge amount of water in their manufacturing process, which we do not see as the end user.
- **Biosphere integrity** – considering how different materials will affect biodiversity in the garden as well as the impact on

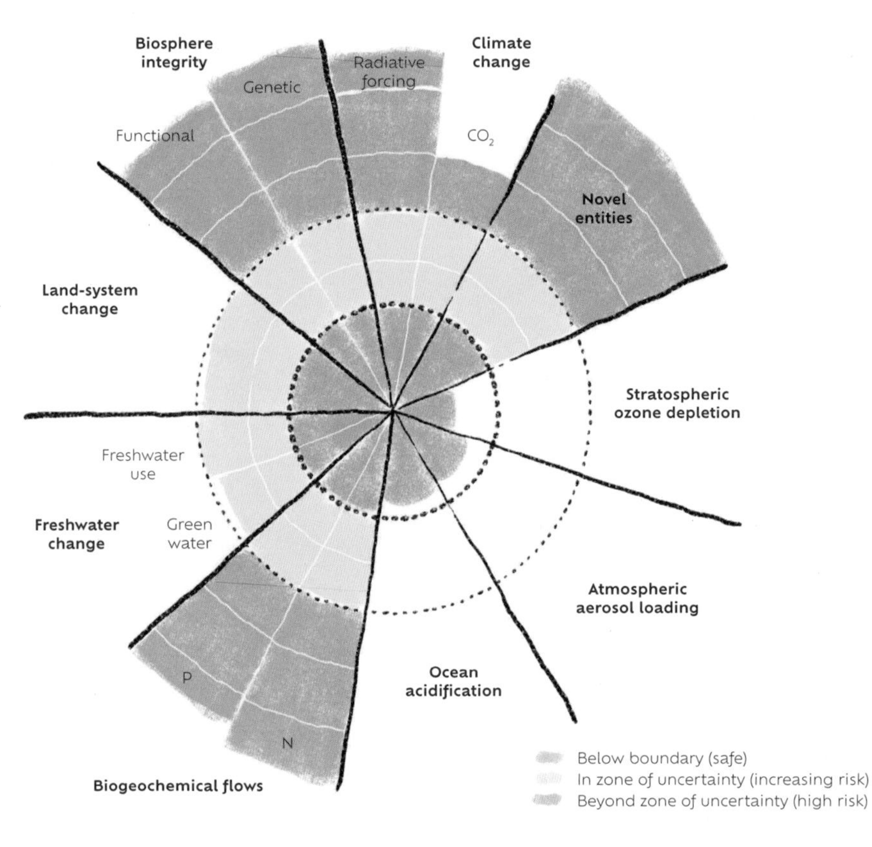

The planetary boundaries framework has useful considerations for decision making, particularly when choosing materials.

There are many different materials available for a variety of budgets, aesthetics, performance requirements, and environmental impacts.

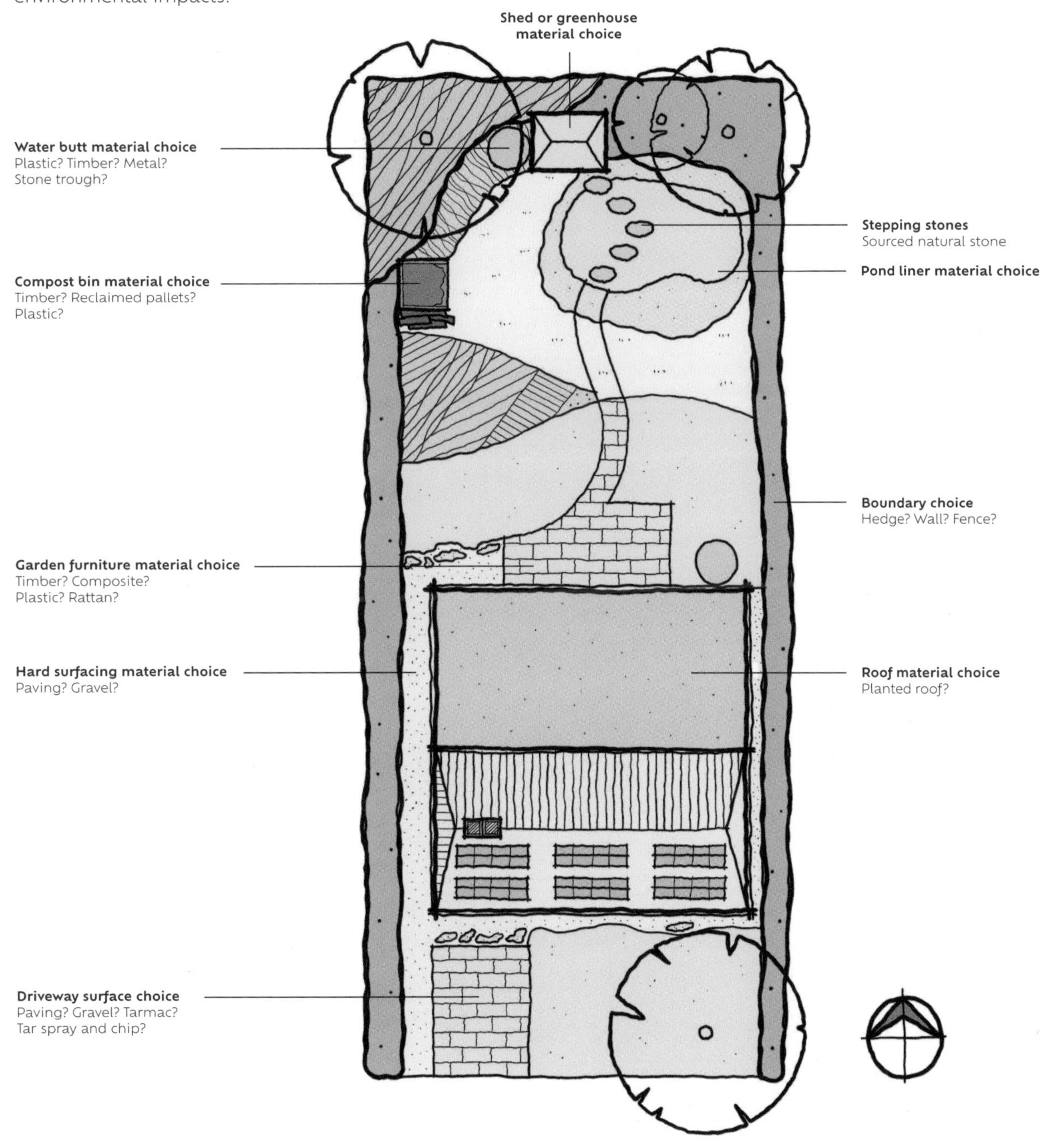

wildlife when manufacturing a product (such as the effect of a quarry or a tarmac plant on the surrounding flora and fauna). A gravel surface, for example, provides more niches for small invertebrates than a tarmac path and will naturally become part of the soil structure as it degrades. However, it is often difficult to measure the total biodiversity impact of a material: for example, although not yet scientifically quantified, there is potential for tarmac to leach chemicals into the surrounding soil, especially in hot weather.

- **Stratospheric ozone depletion** – research the chlorofluorocarbons (CFCs) and other chemicals that destroy ozone high in the stratosphere. Where data is available, look at the relative ozone depletion potential of a material across its lifetime.
- **Atmospheric aerosol loading** – understanding how volatile organic compounds (VOCs) released during the lifetime of a material produce ozone at ground level. Where data is available, look at the photochemical ozone creation potential (POCP) of a material to quantify this.

Environmental Product Declarations (EPDs) present the results of a Life-Cycle Assessment (LCA) of a material or product in a standardised format. A range of environmental impacts are reported, from carbon footprint to water use.

- **Acidification and eutrophication** – looking at measurements of a material's potential to contribute to an overabundance of nitrates, phosphates and other nutrients, which can leach into watercourses and pollute water and the earth.

Assessing all of these factors is complex, and information on the environmental impact of materials is still patchy. Some suppliers are very transparent, while others have very little publicly available data. In many cases it is necessary to make assumptions grounded in an understanding of the material and its use.

Environmental product declarations (EPDs), where available, are excellent sources of information. An EPD presents the results of a life-cycle assessment (LCA) of a material or product in a standardised format. A range of environmental impacts are reported, from carbon footprint to water use.

An LCA evaluates environmental impacts from raw material extraction to finished product, as well as looking at the 'usage' stage and end of life (removal and disposal). It often incorporates a measure of recycling or reuse potential: for example, the carbon emissions offset by reusing rather than re-manufacturing. EPDs are produced following the LCA methodology and are verified by an approved independent verifier before being published.

EPDs can seem overly technical to the nonprofessional, but it is worth downloading them and at least looking at sections A1 to A3, which show the impact of the raw material, transport supply and manufacture of each product and can be used to compare and choose between products. In the table overleaf we have analysed EPDs for some metal, wood, plastic and composite products and illustrated the EPDs in graphic form, adding our own subjective karma rating as a summary. This is a useful process, which I recommend to every design studio and significant user to create a decision-making tool. Some pieces of design software such as Vectorworks and Revit allow data sets to be linked to drawings and have growing databases of the carbon footprint of building materials to integrate in building information modelling (BIM). There are no easily integrated measures of sequestered carbon yet, however, but these will surely come.

In the table below we have analyzed EPDs for some metal, wood, plastic and composite products and illustrated the EPDs in graphic form, adding our own subjective karma rating as a summary.

	NATURAL STONE							
	Limestone	Granite	Slate	Sandstone	Ceramic	Composite Terrazzo	Brick	Low-Carbon Concrete
Recycling or Reuse Potential								
Global Warming								
Water Use								
Ozone Depletion								
Acidification								
Eutrophication								
Photochemical Ozone (Smog)								
Karma Rating								

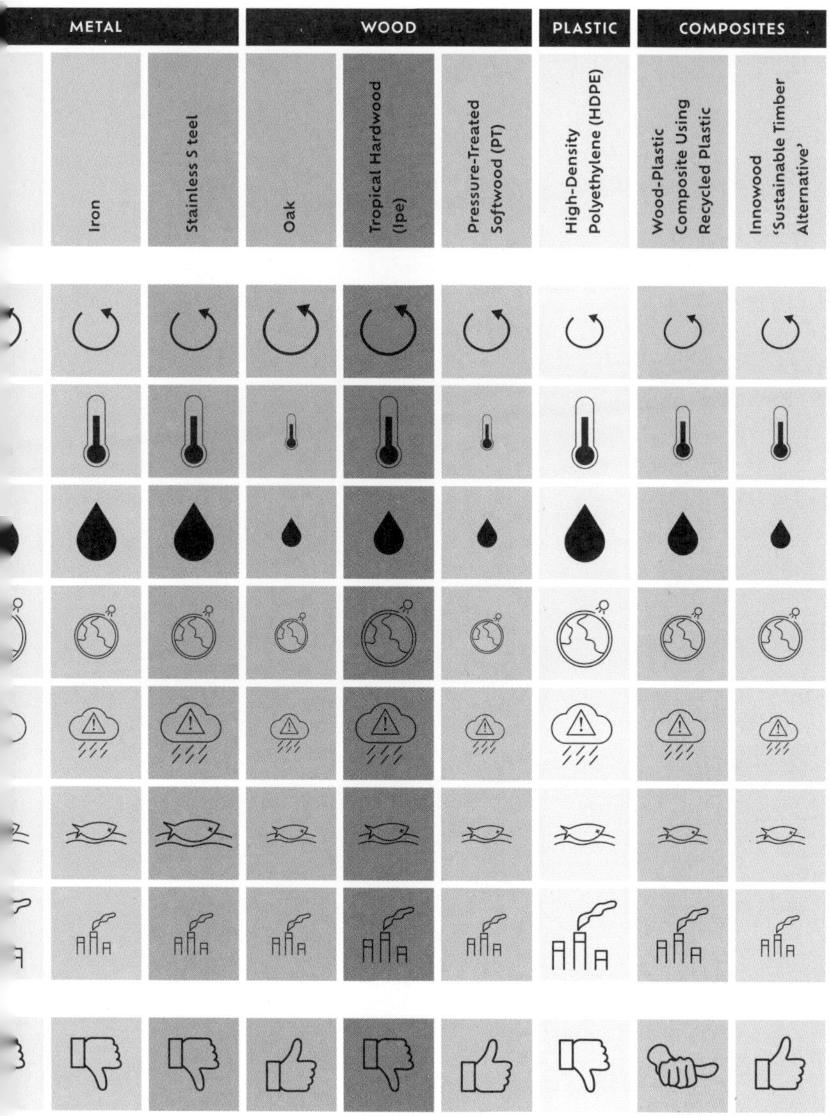

METAL		WOOD			PLASTIC	COMPOSITES	
Iron	Stainless S teel	Oak	Tropical Hardwood (Ipe)	Pressure-Treated Softwood (PT)	High-Density Polyethylene (HDPE)	Wood-Plastic Composite Using Recycled Plastic	Innowood 'Sustainable Timber Alternative'

UNDERSTANDING MATERIALS FOR GARDENS AND LANDSCAPES

> *We learn from our gardens to deal with the most urgent question of the time: how much is enough?*
> *Wendell Berry*

There are so many materials available – how best to choose? When selecting hard landscape materials, it pays to research the local geology through sites like bgs.ac.uk. Visit local quarries, managed woodlands and sawmills. Discover materials that have a relationship with your site, and will meld back into it in time.

Stone

Of the 250 million tonnes of stone quarried annually in the UK, 90 per cent is crushed and used for aggregate. This dense and durable natural material is now being rediscovered by designers. Beautiful, less carbon intensive and typically stronger than brick and concrete, stone is infinitely reusable. It is available in a broad range of colours and textures, but it is heavy, however, so can be costly in fuel carbon to transport. Another consideration is that the thinner it is cut, for cladding for example, the more brittle it becomes, needing more support from other materials.

Limestone is formed from fossils of crustaceans on riverbeds. Its strength, texture and porosity will vary depending on the river water and the shells that form it. Limestone is less hard than granite, which comes from magma formed deep underground, or than marble, which is heated and compressed limestone. Limestone is easier to cut and shape but can be porous and so may stain or weather outdoors, or dissolve in acidic solutions, including polluted rain.

Sandstone is formed of sand compressed over long periods of time. Silica is a very hard material, so is difficult to shape and cut, and the strength will vary depending on how large the particles are and how well the silica is bonded. Yorkstone tends to be very hard, while some other sandstone will weather easily and undergo 'freeze thaw'. This is when water gets inside the natural fissures and blows the layers apart when it expands in freezing conditions.

The key to specifying stone is to ensure it is laid on a loose sub-base or with a lime mortar. If using a cement, ensure it is

Craftsmanship of a new drystone wall with traditional 'hen and chickens' top.

weaker than the stone so can be easily removed for dismantling, allowing infinite reuse.

Wood

This renewable resource sequesters carbon as it grows and holds on to it until it decomposes. It also stimulates an emotional biophilic response in humans. Being with trees is simply good for us. Being extremely versatile, wood fits well in a circular economy as it can be reused several times in different stages and forms, from main timber to smaller pieces or eventually to mulch or pulp. Sustainability is dependent on forestry management. While UK-grown timber requires a felling licence with obligation to restock, 65 per cent of UK timber is imported each year, so certification and supplier traceability are key to sourcing. At the moment most harvested wood is destined for energy production, but applications in construction are growing, particularly with engineered wood. When choosing wood, ask where it is from and if it is certified, whether it has the right strength and durability for the purpose; discover whether excess moisture can be avoided through drainage or by raising it above soil level, and if any

An Alison Crowther seat made of fallen oak. Wood stimulates an emotional biophilic response in humans.

The key to specifying reusable stone is to ensure it is laid on a loose subbase like these sandstone setts, or with lime mortar in walls.

offcuts can be deployed on site or stored to sequester carbon.

The most durable woods for outside are oak (*Quercus*) and cedar (*Cedrus*), although Douglas fir (*Pseudotsuga menziesii*) is a close contender and will be fit for many projects. Slow-growing tropical hardwoods are hard to justify as they take so much longer to replace and are often harvested to the detriment of their environment. The Tropical Forest Foundation, a non-profit organisation committed to environmental stewardship and sustainable forest management, recommends a reduced-impact logging method and offers demonstration models and training curricula in South America, Africa and South-East Asia. Always request certificates before buying wood.

Concrete

Concrete is financially economic, versatile, durable and resilient. It is easy to source, fire-resistant and heat-absorbent, so can be a good material to hold heat and cool areas. Because of the chemical reaction that binds the materials together when Portland cement is mixed with water, however, it also contributes 4–8 per cent of global emissions and is the single biggest building material greenhouse gas emitter.

There are emerging ways of reducing concrete's carbon footprint: by using less Portland cement and substituting ground granulated blast-furnace slag (GGBS), itself in limited supply, or alternatives like limestone fill, calcinated clay and volcanic ash, which are gaining traction. To minimise impact, specify the right strength of concrete for construction, for correct longevity and also deconstruction. Recycle existing concrete on site, or upcycle elements elsewhere. Concrete is heavy on water use in manufacture and construction so substituting rainwater for mains water is also beneficial.

As specifiers, designers and consumers, asking questions when choosing materials is often the first step to minimising our planetary footprint.

HOW TO CHOOSE BETWEEN BOUNDARY MATERIALS: A FENCE, WALL OR HEDGE

In many countries, residential garden fences are anathema, even banned by some US residents' associations. In the UK we tend to have them, perhaps a remnant of our earlier moated ancestors. When driving through the beautiful British countryside, it is a joy to see over hedges and through picket fences into gardens full of hollyhocks (*Alcea rosea*) and ox-eye daisies (*Leucanthemum vulgare*) in summer. It can be a shame when they are hidden behind tall, chemically treated, close-board, softwood fences. Fences like this can have a cheap initial cost and are often chosen to save horizontal space, but they are less good in other ways: too tall for neighbourly chats with a cup of tea, too opaque to see any lurking intruders, and too chemically treated to host wildlife.

The table opposite compares some factors to consider in choosing between a hedge, a wall and a fence. First decide your needs. If security and longevity are the aims, a securely built wall will also provide homes for wildlife and plants in cracks and crevices and hold on to the warmth of the sun. It can be made of reclaimed local stone or brick and be used again if plans change.

Walls retain heat, while untreated fences with gaps can filter wind and be used to grow climbers.

The extra biodiversity of a hedge makes it good for wildlife and foraging. In addition, it soaks up water in wet winters and respires, cooling the environment in a hot summer. It also improves the health of the soil beneath, which is a good donor site for a protozoa tea. Try comparing the managed soil in a garden or field with this more stable state. Dig a hole in both and you should find darker soil with better rhizosheaths on the roots in the hedge line, where no digging or chemicals have interfered.

If you decide on a fence, or there is a fence *in situ*, a good solution is to keep it and put some reinforcing bar mesh up against, then grow lots of forageable climbers up this structure. Community values can be added by sharing them with neighbours – and by putting holes in the bottom for hedgehogs and for neighbours' children to visit.

		Hedge	Fence	Wall
	Wildlife Value	Great for pollinators, birds and small mammals.	Little direct value, but can be a navigation aid for birds and bats.	May provide niches for invertebrates, small plants and climbers.
	Global Warming (Carbon Impact)	Sequesters carbon. One metre of mixed deciduous hedge can absorb 6 kg CO_2 eq per year.	One metre of a typical 2m high timber fence has a carbon footprint of 11 kg CO_2 eq.	One metre of a typical 2m high natural stone wall has a carbon footprint of 1,007 kg CO_2 eq.
	Protection from Pollution	Excellent: intercepts airborne particulate matter (PM) and diverts pollution. Absorbs CO_2.	Some protection – physically blocks certain pollutants but does not enhance air quality.	Some protection – physically blocks certain pollutants but does not enhance air quality.
	Security	Prickly species, such as holly, hawthorn, berberis or pyracantha, form a natural deterrent.	Determined intruders are likely to be able to climb a standard timber fence.	Determined intruders are likely to be able to climb most garden walls without additional protection.
	Wind Protection	Provides a thick yet permeable barrier, which disperses the power of the wind.	Solid fences block wind entirely causing turbulence on the leeward side as the air pressure equalises.	Solid walls block wind entirely causing turbulence on the leeward side as the air pressure equalises.
	Shade and Cooling	Cooling effect through evapotranspiration. Overhanging species can provide shade.	Provides shade, but no additional cooling effects.	Provides shade, but heat absorbed and re-emitted on hot days can have a heating effect.
£	Cost	Starting a hedge from scratch is cost effective. Semi-mature 'instant hedges' are more expensive.	Standard fence panels are significantly cheaper than brick or natural stone.	Most expensive – especially natural stone.
	Time to Reach Full Height	Around 3–7 years to achieve full height. Semi-mature hedges can be planted to provide an 'instant hedge'.	Immediate.	Immediate.
	Maintenance Level	Formative pruning needed in first 2–3 years; maintenance trimming may then be necessary.	Staining and resealing may be required every couple of years. Split and broken pieces will need repairing.	Ageing stone or brickwork may need repointing, but this could be as rarely as once every 50–100 years.
	Longevity	Well-maintained hedges can last for decades. Some hedgerows are nearly 1,000 years old!	Pressure-treated timber fence panels are likely to last 10–15 years.	A well-constructed natural stone or brick wall will last several decades to centuries.
	Durability	More resilient to storm damage than fences, as it slows wind passing through its open structure.	Risk of being blown over or damaged in storms.	Durable in storms, if properly constructed and maintained.

Opposite: This sloped entrance uses asphalt with small angular gravel chippings rolled in, known as 'tar spray and chip'.

Above: The site's own sandstone was too soft to build a retaining wall, but it was perfect for filling retaining gabions.

HOW TO CHOOSE BETWEEN SURFACES

A beautiful entrance sets the tone to a garden and can make you feel welcome and comfortable immediately. Gates, hedges, verges and path surfaces are key. Along with aesthetics, consider longevity, installation and maintenance costs, and suitability for the site of your new surface. For paths consider factors like the frequency and weight of traffic, the steepness of any slopes, inclusive access, the likely temperature, wind and water weathering, the underlying soil and the distance from manufacturer to your site.

Taking these into account, you can then choose the best materials in terms of environmental impact. Sometimes the choice seems intuitively obvious, and sometimes it's helpful to use scientific analysis to help a decision.

In this example I have used EPDs (see page 127) to choose between surfacing a footpath with self-binding gravel or asphalt/tarmac. You can find EPDs for many products and they are generally available from a supplier's website; a quick internet search will give you options. Occasionally EPDs need to be requested directly from a manufacturer, but they should be free of charge. All layers of a material need to be included in the calculations. The typical build-up for these two options is as follows.

- A self-binding gravel footpath would typically consist of 50mm (2in) compacted gravel, laid on top of a 150mm (6in) base of crushed aggregate, with a geotextile sheet beneath.
- An asphalt footpath would typically have a 25mm (1in) 'surface course' of asphalt on top of a 40mm (1½in) 'binder course' of asphalt, laid on a 150mm (6in) base of crushed aggregate and geotextile sheet.

The lowest impact choice may not be the most sustainable if the surface needs to be replaced too regularly. Consider

Above: Cotswold chippings on a permeable subbase is long-lasting and firm underfoot.

Opposite: Some factors to consider when choosing between materials, here for a driveway.

the lifespan of the material and maintenance requirements too. Asphalt typically lasts 15–25 years; however, in very hot weather it can melt and degrade. When self-binding gravel surfaces are repaired, the upper gravel layer is topped up and the aggregate base typically does not need re-laying. In this comparison, I assumed a twenty-year lifespan for the tarmac surface without degradation and that 10 per cent of the surface of the gravel footpath would need repairing every year for twenty years.

When annual maintenance is factored in, the environmental impact of self-binding gravel increases. However, the data shows that it is still a more sustainable choice than asphalt – even over a twenty-year period.

In the chart opposite, the data is shown visually as I find that helps with decision-making.

	REPAIR COTSWOLDS CHIPPINGS Fill potholes with road stone every year for 20 years					RESURFACE TARMAC DRIVE Replace wearing course of driveway once in 20 years				
	Extraction and Processing	Transport to Site	Installation	Disposal	OVERALL	Extraction and Processing	Transport to Site	Installation	Disposal	OVERALL
Global Warming										
kg CO_2 eq	376	111	0	839	**1,326**	2,380	45	205	442	**3,072**
Water Use										
litres	4,580	760	0	2,040	**7,380**	15,470	690	503	1,136	**17,799**
Ozone Depletion										
kg CFC-11 eq	0.00005	0.00004	0.0	0.00016	**0.00024**	0.00110	0.00004	0.00004	0.00008	**0.00126**
Acidification										
kg SO_2 eq	1.5	0.6	0.0	5.1	**7.3**	13.9	0.6	1.5	2.6	**18.7**
Eutrophication										
kg $(PO_4)^{3-}$ eq	0.5	0.2	0.0	1.4	**2.0**	3.5	0.2	0.4	0.7	**4.7**
Photochemical Ozone (Smog)										
kg C_2H_4 eq	0.2	0.1	0.0	0.8	**1.1**	2.5	0.1	0.2	0.4	**3.2**

ENERGY: RENEWABLE ENERGY

> *The garden of the world has no limits except in your mind.*
> Rumi

All living organisms depend on energy that comes from the sun, which is transformed by green plants into energy we can use to eat, wear, build with and power machines. We all rely on this magical transformation of sunlight into energy through plants' ability to photosynthesise. Without it we would not survive. Of the other energy sources – hydro, wind, tidal, geothermal and nuclear – none can directly produce food. Therefore, as people who work with the land, let's concentrate on sunlight first. How can we maximise plants' ability to convert sunlight to energy?

Energy flow

The simple answer is that we need to maximise the energy flow, starting with water and air, as these are the other main factors in photosynthesis. To get good water flow means having enough rain through planting trees and protecting soil. We must ensure nutrient flow through non-compacted soil, allowing healthy underground livestock to keep nutrients cycling by avoiding heavy machinery, poisons and bare soil. We also need to increase the density of green plants to maximise sunlight and carbon capture. We do this by planting in layers, and observing how plants grow naturally: in clusters or guilds, with synergistic root systems.

As growers we can also work to extend the time that plants can grow and the rate they can grow. We can do this by increasing their resources and their access to sun through bigger, healthier leaves. We can give them the right minerals, good access to sunlight, plant them in a good place, create optimum drainage and soil structure with plenty of organic matter and add glass cover to extend the growing season. The mineral cycle, water cycle and biodiversity can all be helped in gardens, woodlands and farmland through management. They can also be disrupted, as our reliance on fertilisers and pesticides has shown, damaging biological communities and lifecycles in the past.

The tools we have at our disposal are: rest, or allowing the land to wild itself gently; natural disruption through fire, flood or human intervention; other living organisms including soil microbes; technology; labour; money to pay for all of these and our own creativity in using them. This last mindset piece is vital and is the key to realising any vision.

Renewables

If we want to use our land to hold and harness other energy, there are several emerging technologies, requiring a wide range of investment and areas of land to be viable.

The first approach is to save energy, or minimise usage, through low embodied carbon design and building, upcycling existing structures and substructures and recycling materials. We can specify timber frames in construction from locally managed woodlands, locally quarried stone, natural insulation for buildings and glazing optimised for whole-life carbon emissions. Rainwater can be harvested for any irrigation or water features and for flushing loos if viable. Low-flow water fittings can be installed to reduce usage too. Any outdoor lights can be solar powered and pointed to the ground or incorporate bat hats to minimise disruption to wildlife.

Air source heat pumps are relatively easy to retrofit on a south-facing wall or outbuilding, and can be powered by photovoltaics to access warmer midday and afternoon ambient air outside, to transfer to underfloor heating, radiators or a swimming pool. Heat pumps use refrigerant, which could leak and emit greenhouse gases, so it is important to specify innovative air source heat pumps that use low global warming potential refrigerants.

A water source heat pump converts heat energy from water to provide heating and hot water. The image oveleaf shows a

Of the many potential energy sources we can harness on a larger scale, the most available to gardeners is sunlight.

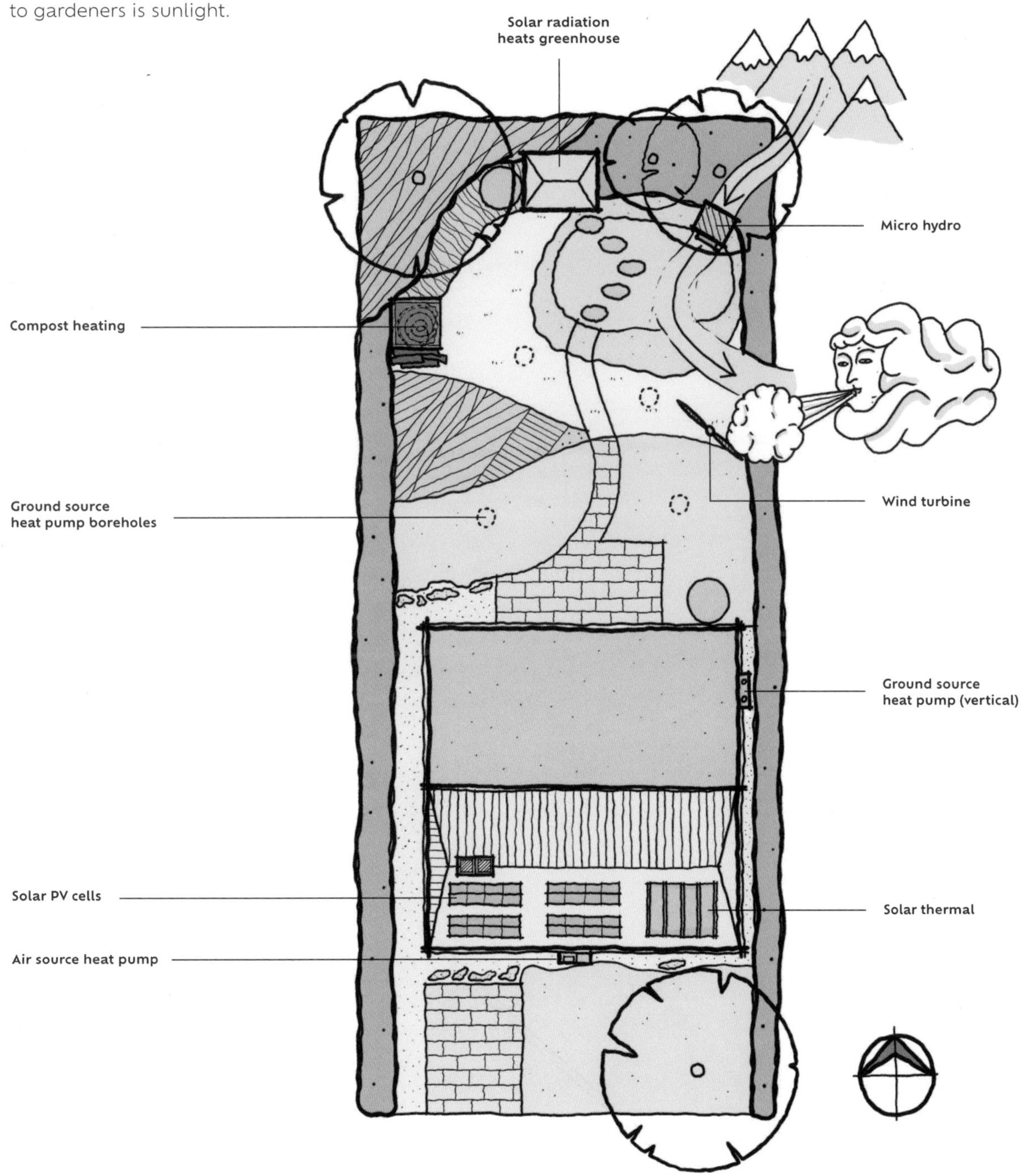

Solar radiation heats greenhouse

Micro hydro

Compost heating

Wind turbine

Ground source heat pump boreholes

Ground source heat pump (vertical)

Solar PV cells

Solar thermal

Air source heat pump

A water source heat pump converts heat energy from water to provide heating and hot water.

newly dug 0.8-hectare (2-acre) lake my studio designed to have water source heat loops below the surface, which provide heat to a Grade II listed house, as well as creating new habitat for rare flora, including the nationally scarce white-clawed crayfish. Water source loops need to be covered with at least 1m (3ft) of water at all times.

Ground source heat pumps take geothermal energy stored in the form of heat in the rocks and soils beneath the surface of the Earth and use it to heat our homes. They need electricity and reduce the underground temperature around them slightly, to convert approximately 1kWh of electricity to 3kWh heat. Ground source heat pumps can be laid in horizontal loops or drilled vertically down in boreholes up to 150m (500ft) deep. Vertical loops will impact planting less as soil surface temperature is less affected, and the network can also generate cooling in summer, which recharges the boreholes for winter. Multiple boreholes will be needed depending on the amount of heat required and should be placed about 10m (33ft) apart. Thus, while they can be fitted in a large garden, a spare paddock or field is handy.

Solar photovoltaics convert the sun's energy to electricity via the photoelectric effect. This is non-mechanical and there are no emissions or noise, so solar photovoltaics can be mounted on roofs or free-standing on the ground. Optimum positioning is a south-facing and shadow-free roof. It is worth looking for an environmental certification such as Cradle to Cradle, which ensures materials are recycled at end of life and minimises embodied carbon, which varies hugely between products. Some solar photovoltaics also offer a forty-year warranty and performance guarantees. Ground-mounted solar should be installed on low-grade farmland (grade 3b or below) away from flood risk, and can be planted with wild flowers and grazed by sheep. Studies have shown that the cooler air from surrounding planting can improve the solar performance by around 5 per cent. Depending on local access to the national grid, electricity can be shared or sold back as well.

Biomass boilers take up a lot of space and are good to incorporate in a woodland management regime for a small estate, but have lost popularity since the UK government renewable heat incentive was withdrawn in 2022. Woodchip for the biomass boilers needs to be the right moisture level and can be carbon intensive if transported from afar or even imported from abroad as some UK woodchip is. A wood-burning stove, where permitted, is a good way to supplement more passive energy sources in very cold weather, and can be designed to be extremely efficient and supplied by sustainable, locally managed woodlands.

Hydropower harnesses the energy of fast-moving or falling water and is feasible as a balance for solar in some UK areas, when heavy winter rains can cause faster flows just as solar is reduced. The drawback in most cases is the high volume of water flow required and the embodied carbon, cost and space of the infrastructure. There are also some environmental impacts that need to be mitigated: to allow fish runs, for example. For sites on steep hills in rainy locations or by tidal flows, hydropower could be considered, but it is less helpful for most residential sites.

For a wind turbine to be feasible, studies show that average wind speeds at hub height should be higher than 5.5m (18ft) per second, since the power generated is proportional to the wind speed. A wind turbine would usually be about 25m (82ft) high with a blade diameter of 19m (63ft) to achieve this. It would also need to be located at a distance of ten blade diameters away from homes, because of the noise. Therefore, it is not feasible for most homes under current technology and policy, but there are some hopeful designs coming from the Netherlands, for example, that may be suitable for smaller sites in the future.

These last two technologies (wind and hydro), along with other sources of generating electricity like nuclear power, are best accessed through facilities producing them at scale. Some countries such as France, which is almost fully nuclear powered, have closed their fuel cycle by recycling, reprocessing and reusing their spent fuel. Other countries are further behind. Much of this is policy-led, so as land stewards and gardeners our best contribution is to ask questions, understand our sources and options, and get involved, so that we will be able to continue to access abundant energy in a more circular economy in the future.

HOW TO WORK WITH NATURAL ENERGY

> *Everything is energy and that is all there is to it.*
> *Albert Einstein*

Once we understand some of the energy that we have available for free, our gardens can benefit from these natural resources and we can garden with the flow, buying fewer inputs and making less effort for the same or better results.

Sun

Too much or too little sun can be make or break in a garden. As part of maximising our plants' ability to photosynthesise, we can design our gardens to capture as much of the sun's energy as possible, and also mitigate against excess heat. Extend the use of stored heat at the edges of the day and seasons by growing under glass, or by growing against a south-facing wall to protect from frost and help fruit ripen earlier. Position trees to provide shade for a hot day, and create different ecosystems in a garden. Trees and ponds are also air conditioners. Plant deciduous trees on the sunny side of a building: low winter sun will provide light and warmth through bare branches; when in leaf they will shade buildings from summer sun. A deciduous climber works too, and on cooler aspects will help insulate and protect a building from prevailing weather.

Wind

This is one of the most powerful energies in nature. Because the sun's energy heats the Earth more at the equator than the poles, and the Earth's rotation usually makes the ocean's currents move from west to east, our prevailing wind in the northern hemisphere is the south-westerly. This means the wind and the weather generally come from that direction, explaining why we can track rain coming our way. Hills and forests will intercept rain on the way, so there is often less rain in the east of the UK. North-easterlies, although rarer, bring colder air, so prepare for them by planting sensitive plants in the lee of shelter belts. Hedges also provide shade and shelter. They act as a filter, slowing the wind but not creating the solid barrier of a wall, which produces eddies on the lee side as incoming wind hits it

Once we understand some of the energy that we have available for free, we can garden with the flow.

We can design our gardens to capture as much of the sun's energy as possible, and also mitigate against excess heat.

and curves over the top. Trees are an excellent wind filter and have the benefit of sending wind up high over their tops, as well as filtering it. Even a small tree can be a big benefit to a building or garden. Plant copses and trees in parkland down a hill or further away from a larger site, as the effect will be felt for a distance of multiple tree heights. There are advanced wind maps available online, and it's worth understanding weather patterns to understand the context of your site.

Gravity

Gravity affects wind, steadies the Earth in orbit around the sun, and holds the moon in orbit around the Earth. The moon takes 27.3 days to go around the Earth and rotates at the same time, in such perfect sync with Earth that we never see its 'dark side'. The moon stabilises the Earth's tilt, which causes our seasons, and the gravitational pull of the moon and the sun cause our tides. We garden according to the seasons caused by orbiting around the sun, and the moon's gravitational pull can likewise be used to plan the gardening month. Like indigenous wisdom keepers, biodynamic and permaculture practitioners, we can use the moon cycles to maximise yields. When the moon is waning, drawing the energy down into the Earth, sow seeds and work with root crops. Deal with plants above ground when the light is strongest and energy greatest. Maria Thun's calendar is still one of the best to follow, and the chart on page 212 is an adaptation of her teachings. The results will be amazing especially when combined with living compost. However, in the spirit of the kindest garden, follow Christopher Lloyd's maxim: the best time to do something is when you should, and the second best is when you can.

Water

Life, movement and energy are brought to a landscape by water. It connects the sky to the Earth and plays with light and shadow. Plan moving water and still water through your site to regulate heat through evaporation and condensation. It will cool you in summer and hold some groundwater heat in winter. It is vital to all soil interactions and photosynthesis.

Masaru Emoto and Viktor Schauberger have done fascinating work on the structure of water and described water's ability

Trees are an excellent wind filter and have the benefit of sending wind up high over their tops.

Land energy crossing points are good places for siting items that will help to ground, or will benefit from, a flow of energy.

to absorb and reflect the energetic vibrations it encounters. Humans are over 70 per cent water, so we are major repositories of these vibrations.

Install a biodynamic flow-form water feature to oxygenate, structure and energise water; on a small scale, a compost tea bubbler or handheld drill energises too (see Making a compost extract, page 63). Add humates to mains water to remove chloramine before use on plants and filter before drinking.

Magnetism

The Earth has a magnetic field generated by its core. In the last fifty years scientists have discovered how magnetic fields are used by migrating birds, sea turtles and salmon to find their routes, and by some birds to orient their nests. Energetic fields are also thought to have informed the positioning of the pyramids, stone circles and sacred land sites for millennia, while in 2018 physicists showed that, in the resonant state, the Great Pyramid of Giza, Egypt, can concentrate electromagnetic energy. Since the nineteenth century the influence of this energy has been explored in the West to try to explain how it might influence everything from the human immune system to where insects like to congregate and the best position for a compost heap.

Biodynamic farming acknowledges the dynamic forces in nature not yet fully understood by science. By working creatively with these subtle energies, regenerative farmers and growers can enhance the health of their farms and the quality and flavour of food. In a garden context try using dowsing rods to locate underground water or the best place for a compost heap. Tune in to land energy by sitting quietly and practising energetic listening to locate the best position for a building, a standing stone, a healing tree or a labyrinth.

HOW TO WORK WITH LAND ENERGY

> *When you touch one thing with deep awareness, you touch everything.*
> *Thích Nhất Hạnh*

The subtle energies of the land are not fully understood, and although utilised over millennia by farmers, feng shui experts and Earth energy healers, it's an area easily dismissed. Over the last century a new generation of scientists has begun to operate across the boundaries of biology, chemistry and quantum physics to pioneer research in an area called quantum biology, investigating how phenomena such as photosynthesis, respiration, bird navigation and even the way we think are all influenced by quantum mechanics. Quantum refers to the dimension that is subatomic, so they are talking about things that work on an energetic level, via frequencies, waves and photons. By keeping an open mind, regenerative growers are now using observed outcomes without waiting to understand all the science.

We can all use this new interest in ancient communication to enhance the health of our land and the quality and flavour of our food, by observing the effects in gardens and being informed by the results in plants and soil. This will no doubt be the subject of many books to come, so do follow up some of the links on pages 233–4, and in a garden context try some of these techniques for yourself.

Paramagnetism

The work of pioneers like Dr Phil Callahan has shown that, depending on the configuration of their electrons, materials can be magnetic in different ways and to differing extents. Paramagnetic trace elements can produce a measurable magnetic charge in some very old volcanic rock dusts. This slight charge increases the amounts of nutrients available to plants, which are diamagnetic. Try adding commercially available volcanic rock dust to your vegetable garden soil in some areas; then observe the difference in Brix measurement (see page 199), pest resistance, appearance and taste – the best assessment of metabolites providing nutrient density.

If you are curious about paramagnetism, try the less-often explained tools described below.

The subtle energies of the land are not fully understood although recognised over millennia at points such as this standing stone.

Above: Try using dowsing rods to locate underground water.

Opposite above: By keeping an open mind, regenerative growers are now using observed outcomes without waiting to understand all the science.

Opposite below: Combine energetic listening with dowsing to site things like seats, vegetable gardens, labyrinths and standing stones.

Dowsing for water

Many farmers and water engineers find dowsing rods are still the most efficient way to locate the gentle magnetic fields generated by flowing water beneath the soil. You too could experiment by trying to find water with dowsing rods. It is a good idea to do this at a known water source the first time, so you get used to the feeling of the rods moving.

Take two copper dowsing rods and hold them firmly but gently, one in each hand. Focus on water. Walk slowly in a straight line with the rods held at elbow height, your elbows tucked comfortably in to your sides. Note where you are when the rods begin to cross and keep walking until they uncross. Repeat at right angles to pinpoint a specific spot. Advanced dowsers can sense the depth of the water by saying or thinking of depths in increments and seeing when the rods cross. If that piqued your interest, move on to choosing the best place for a compost heap, standing stone or labyrinth.

Dowsing for an energy spot for a compost heap

Land energy crossing points are good places for siting items that will help to ground, or will benefit from, a flow of energy. This could be a good position for the centre of a labyrinth or a standing stone, for example. A flow of energy is also thought to draw insects and creatures in, so could be suitable for a compost heap in a practical setting.

You can use dowsing rods to find these subtle energy points in the same way as water. Cross the area at several points to find a useful intersection, and build your compost heap there. Obvious caveats apply on balancing the optimum energy point with practicalities like ease of access and safety for your compost heap.

Energetic listening

This is an extension of the above, using our own body as a sort of antenna or intuitive listening instrument for creative co-working with the land.

- Sit on the ground, on a log, on a rock or with your back to a tree.
- Set your intention to resolve a specific issue or just to be curious about the space.
- Clear your mind through breathing techniques or by connecting to the land in your imagination.
- Try not to think or make up stories in your thoughts.
- If nothing happens that is fine too.
- Close your eyes and listen.
- Start with listening to the sounds. You might count them to stop your mind wandering.
- Extend your listening to feeling your own body from within. Gradually extend to feeling beyond your body. Feel the grass, the plants, the trees, the water, the land and the site.
- Listen to what the place is 'saying'.
- How does it feel, what does it need?
- Try to be just curious; listen without judging or jumping to solutions.
- After a few minutes open your eyes and write or draw.
- Then look around at the site again and see it within the context of what you 'heard'.
- Judgements, rational mindset, solutions and questions will soon follow.

Combine energetic listening with dowsing to site things like seats, vegetable gardens, labyrinths and standing stones. First listen to what the land would like, then find the optimum location. If you follow it through, see if you can tell if your experiment worked. With many curious minds we can advance our observations and better growing.

RESILIENT PLANTING

> *Resilience is the capacity to adapt with grace to the circumstances of life.*
> *Thích Nhất Hạnh*

Resilience means adapting our gardens and landscapes to survive better through extreme wet and dry weather. When added together, our millions of small interventions could also rebalance some of the effects of these climatic shifts. Planting for future resilience is tricky when we don't know exactly how the climate will change, yet with uncertainty comes a certain freedom. Our gardens become an experimental space. We can learn to be curious and responsive rather than chasing a notion of perfection.

Current predictions suggest that London may become similar to northern France by 2030 and to Barcelona by 2050, which means heavier rain and more intense heat for most of the UK. You should therefore plan for these two extremes in your garden.

In torrential rain, where does the water flow? Video where rain collects in rainstorms (this also highlights broken gutters). Heavy rain causes soil compaction, impeding root growth, causing soil stagnation and run-off. The first step, then, is to slow the flow from the sky to the soil. Plan tree canopies, shrub layers, multilayer herbaceous and ground cover, and dress pergolas and walls with climbers to intercept rainfall. Choose plants that will prevent soil erosion with their root structures; mulch your soil and feed your soil microbes to increase the soil sponge effect.

The second scenario is extreme or prolonged heat, when healthy soil and shaded areas become key. In heat spots plant resilient trees, hedges, climbers, shrubs and ground cover first, saving perennial herbaceous layers for later phases.

The aim with both imagined scenarios is to return towards a temperate state, to add grades of layered shelter and to hold a good balance of moisture and nutrients in the soil, the plants, and at different heights and depths to extend the functioning ecosystems. These layers will also help with extreme wind, preventing moisture loss through transpiration by filtering, slowing and absorbing wind, whereas close buildings and walls can often do the opposite – adding turbulence and speed by channelling the wind into smaller spaces.

Once you have started addressing soil health, plan for the plants that suit your conditions. Understanding that potentially 30–50 per cent of each plant is visible above ground underscores the significance of robust root systems; resilient plants tend to have a taproot, fibrous side roots or both. These roots act as an anchor, seeking water at different soil depths, storing energy, sequestering carbon and nurturing symbiotic relationships with soil microbes, giving exudates in exchange for minerals and nutrients from the soil. If the plant is a legume it should

When added together, our millions of small interventions could help rebalance some of the effects of climatic shifts.

A small garden will not feature all planting types, but here are a few areas that could be inspired by natural archetypes.

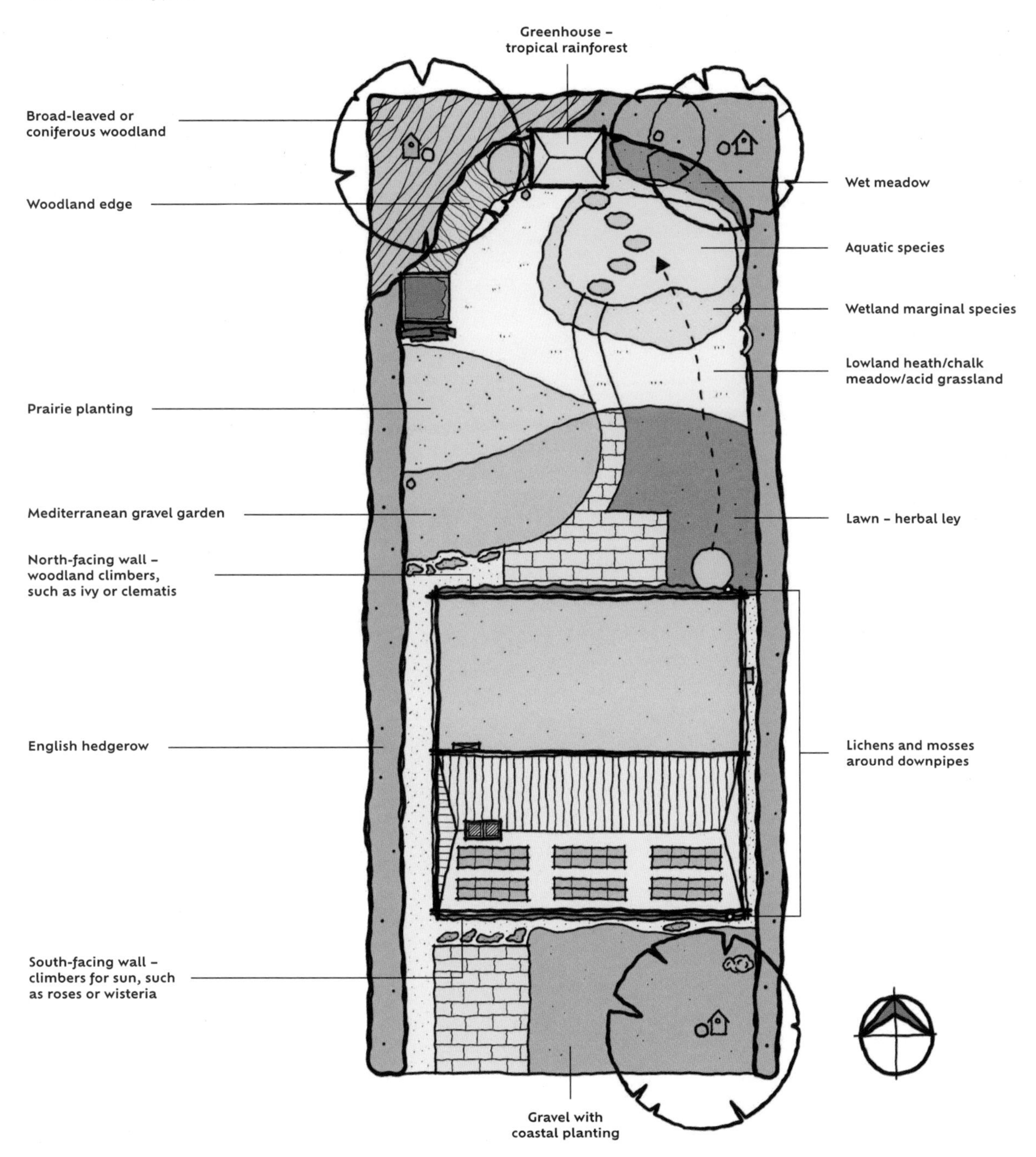

Like humans, plants need resilience in community to survive tough times.

fix nitrogen too. Those roots and their mycorrhizal fungi are essential for holding soil together in heavy rain, and as they decompose they create channels for water to reach deep aquifers. Therefore, when you're harvesting or weeding, try to keep some roots in the soil by hoeing, cutting back or trampling rather than by pulling or digging.

Like humans, plants need resilience in community to survive tough times. Create this by planting in guilds. A guild is a planted community – a mutually supportive group of plants that provides functional diversity by mimicking a naturally occurring archetype: a woodland edge, a meadow, a wetland or a prairie, with layers of habitat, protection and nutrition.

When choosing plants, consider where they originate from and how their favoured conditions match your local site. Most plant guides will note preferred soil texture, moisture, pH, aspect and sunlight. To this, add your own notes on where they thrive, by latitude, altitude, inland or seaside, the side of a road or middle of a wood, and so on, and with which plants they have reciprocal relationships, in their natural habitat. Without becoming too academic or list-driven, we can get a feel for communities and locations with similar characteristics whether in the countryside or in a city. The plants that thrive in those situations will suggest both a palette to choose from and lessons in adaptation.

These larger-scale archetypes have useful lessons to apply to smaller ecosystem creations. When starting from scratch you can visualise a design by sketching over photographs to observe natural distribution patterns and to create a layout plan, an elevation sketch to show how layers work together, and finally your planting plans or matrices to make best use of the space.

With the limited space available in this book, I have included one example of each of these methods for four archetypes, to use as inspiration when creating communities. A small garden would not include them all, of course, but the diagram on page 161 shows some areas that they could apply to.

LARGE-SCALE WOODLANDS: INSPIRATION FOR OUR GARDENS

> *In a forest, a tree is a tree, but when you look deeply into the tree, you can see the sunshine, the rain, the clouds, the Earth, and the minerals. You can see the entire cosmos in the tree. The tree and the cosmos inter-are.*
> *Thích Nhất Hạnh*

The woodland network is an extraordinary ecosystem of interconnected life forms. At its core are the trees, which provide the shelter above and below ground. Trees communicate through a mycorrhizal network between roots: a huge network of fungal threads interconnecting same and different species, exchanging information and nutrients through chemical messaging. Trees might look like individuals, but they are connected below ground to a mother tree; some aspen (*Populus tremula*) forests have been shown to be one vast connected tree! The mother tree feeds, teaches survival strategies and provides 'instructive shade' to regulate young trees' growth, preventing them growing too tall and leggy.

Trees also adapt to changes in climate. Through epigenetics they 'remember' the past year's weather and consume less water following a drought, or grow more slowly in times of scarcity. Scientist and forester Peter Wohlleben suggests the lack of a mother tree in field-grown transplants can explain why young trees might make 'strategic mistakes' like dropping leaves too early in temporary summer dry patches. Without a mother tree to send signals they must learn on their own.

In case of sickness a tree may die back but reshoot elsewhere – suckers are a sure sign of stress, but also of hope for regeneration. If this happens in your garden you can select the best sucker to be the future generation. As well as feeding soil microbes with exudates, trees feed each other, providing nutrients to sick or old stumps like an old matriarch. Close relationships can form between species: for example, beech (*Fagus*) and ash (*Fraxinus*) intertwine stems and roots but avoid crossing canopies, which would block each other's sunlight. When one of these dies, the other soon follows.

Professor Monica Gagliano's work shows trees' ability to hear sound: they react to insect attack and even the vibrations of a chainsaw with chemical pheromones, which warn others

Create ecosystems made up of layers like woodlands, to encourage the natural communication and energy flows above and below ground.

The mother tree feeds, teaches survival strategies and provides 'instructive shade' to regulate young trees' growth, preventing them growing too tall and leggy.

of attack. Wohlleben has shown how wattle (*Acacia*) trees use airborne pheromones to warn other trees of giraffes browsing, causing them to make their leaves bitter. Of course, clever giraffes have learned to browse upwind!

Trees also create rain. In the UK and Ireland, for example, where around 37 per cent of the land area has the potential for reforestation, it is estimated that doing this would increase precipitation by an average of 0.74mm ($\frac{1}{35}$in) per day (24 per cent) in winter and 0.48mm ($\frac{1}{50}$in) per day (19 per cent) during summer. Several factors potentially contribute to trees' ability to make it rain. Forests and woodlands have a higher surface 'roughness' than agricultural land. This produces more turbulence over the trees and slows the movement of heavy clouds, causing them to rain over and downwind of wooded areas. The same is true of urban areas, however: increased surface roughness from buildings can cause heavy rain over cities and downwind of them. Therefore the cities may harvest the rain, but woodlands originate it, since they evaporate more water than both cities and most agricultural land, particularly during the summer season. The water that evaporates from their leaves creates a microclimate, forming an area of low pressure above the tree canopy, contributing to what is known as the rain corridor stretching between forests.

Temperate woodlands are havens for wildlife. Although woodland trees are mostly pollinated by wind, birds and insects are vital parts of the woodland network, as they disperse seeds and fruit. The best way to start a new woodland is to leave little piles of acorns out for the jays, who will take and bury them for winter. The ones they don't eat will grow. Birds also help ensure insect populations stay in balance. The insects on the forest floor are the vital clean-up crew. Detritivores like beetles, woodlice and soil dwellers like springtails break down leaves and rotten wood to allow fungi and bacteria to recycle them into nutritious humus.

Around 40 per cent of the UK's ancient woodland was replanted with dense single-species conifers in the 1940s and 1950s for commercial timber production. These barren monocultures lack vital diversity of root structure. Many have been toppled in recent storms, while others are being felled for native woodlands to be restored. The effects are wonderful: by allowing gradual light and diverse layers of ground cover, subshrubs, deadwood and successional growth to recolonise, species like the purple emperor butterfly, lungwort lichens and wood anemones (*Anemone nemorosa*) are returning along with songbirds and rare woodland molluscs.

Allow similar ecosystems of layers to develop in your garden to encourage the natural communication and energy flows above and below ground that we need to regenerate living landscapes. Tune in to the effect of a contented woodland on your own nervous system, to the feel-good hormones like serotonin released by our bodies when among trees. Breathe in deeply to inhale the oxygen, phytoncides, antifungal and antibacterial chemicals released by trees, which strengthen our immune, hormonal, circulatory and nervous systems. Breathe out to share your breath with the trees.

Tune in to the effect of a contented woodland on your own nervous system to recreate it in your garden.

HOW TO PLANT A WOODLAND EDGE, HEDGEROW AND MINI FORAGED WOODLAND

The edges of things are where magic happens, with more diversity concentrated in these liminal spaces than in dense woodland or open grassland. The hedgerow and woodland edge are iconic looks as well as vital corridors and refuges for wildlife in the UK countryside. In a garden context we can recreate this multi-layered ecosystem not only with hedgerow and woodland-edge plants but also as an archetype to inspire successful mixed herbaceous borders.

Hedgerows

Start with a hedge to provide a backdrop to a planting scheme. Include forage and berries for birds, especially migrating visitors like redstarts, which rely on our hedgerows for sustenance after their long flight from Russia in winter. Plant small whips in staggered rows, and if you have rabbits use biodegradable rabbit guards to give plants a head start. Mulch with a thick layer of broad-leaved woodchip.

Woodland edges

Natural woodlands vary in character depending on location, soils and age, but will typically be made up of tall standard trees like ash (*Fraxinus*), oak (*Quercus*), lime (*Tilia*) or beech (*Fagus*) in the UK. They support huge numbers of invertebrates in rich understoreys among mosses and lichens, ferns and fungi.

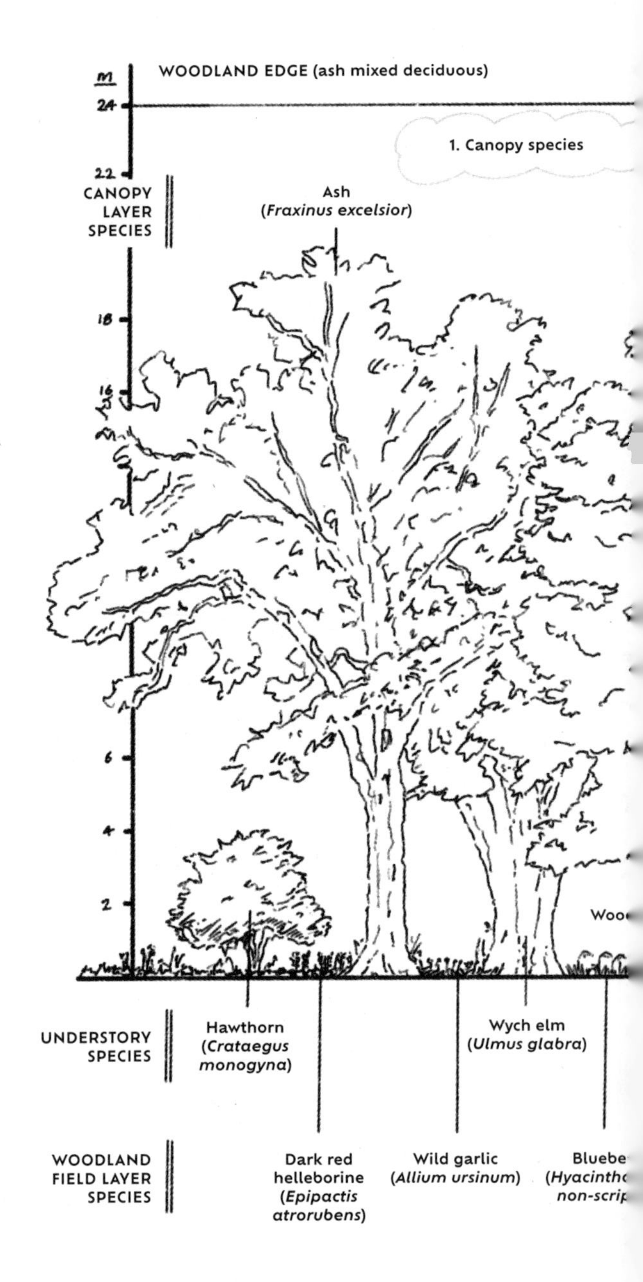

The hedgerow and woodland edge are iconic looks as well as vital corridors and refuges for wildlife in the UK countryside.

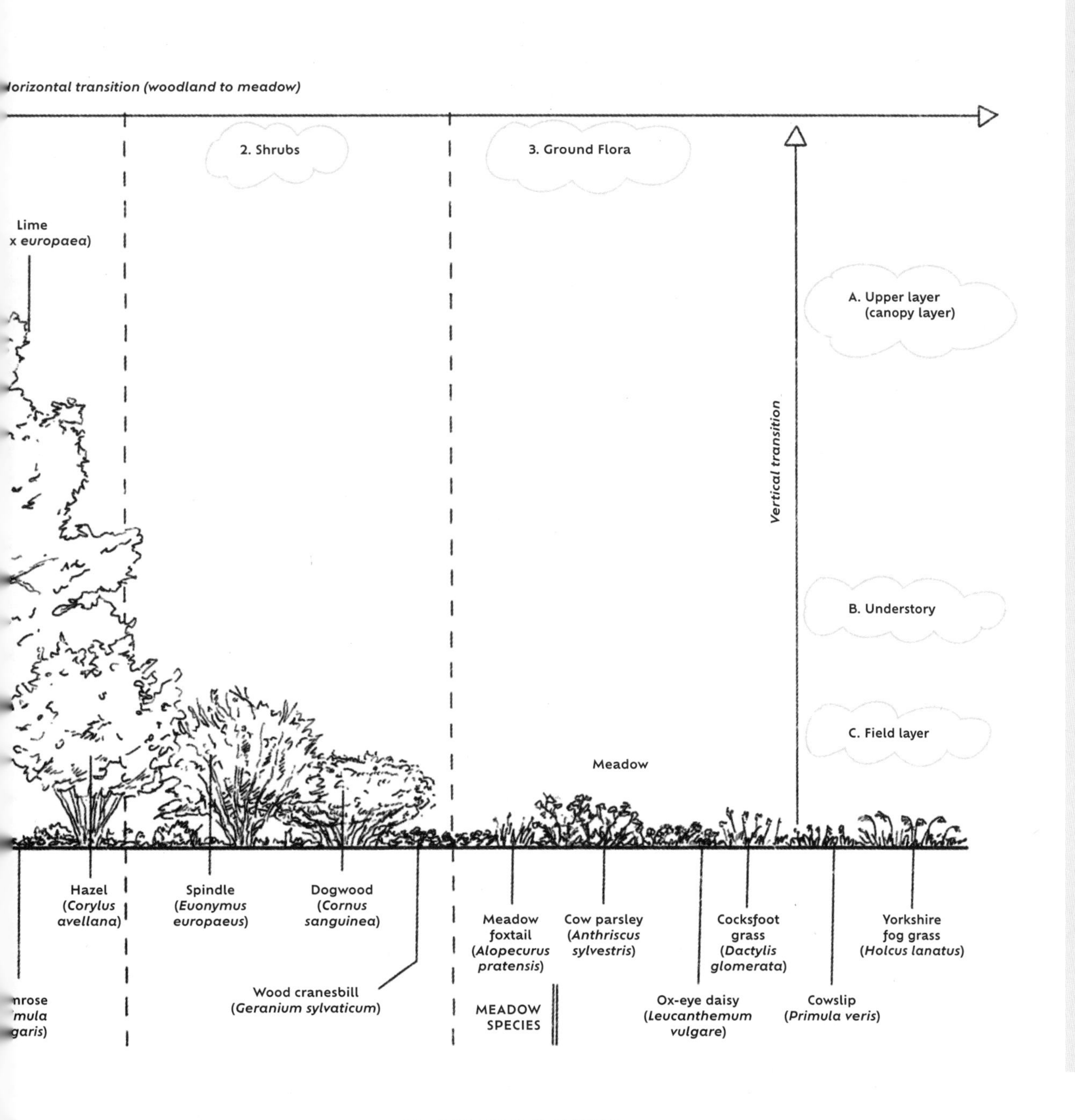

When an old tree falls or is coppiced, colonisers like birch (*Betula*) may pop up in the new patch of light, along with bulbs and corms like wood anemones (*Anemone nemorosa*) and bluebells (*Hyacinthoides non-scripta*), and grasses like melic (*Melica*). These flower in spring and then go dormant as the tree and shrub canopy fills out in early summer. Towards the edge of the woodland, primroses (*Primula vulgaris*) and violets (*Viola*) will grow; the additional sunlight also encourages shrubs like hazel (*Corylus*), spindle (*Euonymus europaeus*) and dogwood (*Cornus*). On the ground, wood cranesbill (*Geranium sylvaticum*), cow parsley (*Anthriscus sylvestris*), ox-eye daisy (*Leucanthemum vulgare*) and grasses like cock's foot (*Dactylis glomerata*) and Yorkshire fog (*Holcus lanatus*) blend together, with a variety of heights, spread and roots covering different soil strata.

Translating woodland-edge plants to a resilient garden palette means taking their characteristics and using natives or near natives to fill the different roles and layers in an archetype or guild.

Trees

In a small space you might add a single tall tree above a medium shrub layer to host maximum species and intercept most rain and sunshine. Choose the tree depending on your soil, location and the amount of shade you would like. Woodland-edge planting needs a more fungal soil, so if you are starting from scratch add a living compost and plant through that, particularly near the shrubs and trees.

In a garden with clay soil, choose an ornamental pear like the upright *Pyrus calleryana* 'Chanticleer' or crab apple (*Malus*) 'Evereste' to provide shade and forage. A hornbeam (*Carpinus betulus*) will provide a deep shade canopy or, in a larger site, choose small-leaved lime (*Tilia cordata*) for its scent and edible leaves and flowers.

Mid-layers

The beautiful berried spindle (*Euonymus europaeus*), cornelian cherry (*Cornus mas*) with the earliest flowers and edible fruits, bridal wreath (*Spiraea arguta*) and the flowering dogwood *Cornus kousa* from China will form a thriving mid-tier in a garden. For a wilder feel and forage, choose blackthorn (*Prunus spinosa*), common elder (*Sambucus nigra*) and sweet briar (*Rosa rubiginosa*) for flowers, berries and hips.

Understoreys

Choose layers of taller foxgloves (*Digitalis purpurea* or *D. parviflora*) with showy toothwort (*Cardamine pentaphyllos*) and barrenwort (*Epimedium*) below. With a little more sunlight, perhaps on the south-facing side of a copse of trees, peonies (*Paeonia*), hellebores (*Helleborus*), columbine (*Aquilegia*) and cranesbills like bloody cranesbill (*Geranium sanguineum*) will provide the emergent layer with spring-flowering and summer-dormant wild blue phlox (*Phlox divaricata*), to be followed by autumn-flowering white wood aster (*Eurybia divaricata*). Ground cover like lungwort (*Pulmonaria*) and bugle (*Ajuga*) and summer-dormant bulbs like lily of the valley (*Convallaria majalis*) can pop through below. In heavily shaded areas, ferns like shield fern (*Polystichum*) and the very tough ground ivy (*Glechoma hederacea*) provide shelter for small mammals and invertebrates and protect the soil. Not all of these plants are native to the UK, although many have naturalised here or been grown for hundreds of years, like cornelian cherry (*Cornus mas*), which is first recorded at Hampton Court Palace gardens in 1511.

The aim is to provide many layers of diverse habitat, different nectars and forage types, so mix natives and non-natives to provide the widest range of food. Keep more exotic plants nearer the house and concentrate on natives towards the boundaries. This helps the garden edge harmonise with the wider landscape, and prevents garden plants from colonising the wider landscape.

Mini forage woodlands

For a small garden guild, use a fruit tree, for example a self-fertile apple (such as *Malus domestica* 'Egremont Russet') on an M26 (semi-dwarfing) rootstock as a main structural element, providing shade, food and shelter for many creatures. Below this tree layer create a nourishing ecosystem of nitrogen-fixing shrubs like autumn olive (*Elaeagnus umbellata*) or sea buckthorn (*Hippophae rhamnoides*) and dynamic accumulator emergents like comfrey (*Symphytum officinale*) or white lupins (*Lupinus albus*), which have taproots that can access nutrients from deep inside the ground and make them available to other plants. Add ground cover and herbaceous layers including plants with fruits or autumn seed heads for forage.

Choose annuals like sunflowers (*Helianthus annuus*), love-in-a-mist (*Nigella*) and poppies (*Papaver*) to help establishment. They will provide weed-suppressing ground cover and contribute to the all-important compost layer as their whole body mass dies each year after setting seed. As they become shaded out by the growing tree and shrubs, they will fade out of the palette, but any seeds you don't harvest will remain in the soil to compete with other weeds if the ground is disturbed.

MEADOWS

Meadows store carbon, slow down water, reduce flooding and pollution, and provide excellent habitat for pollinators, small invertebrates and mammals. They also bring great joy. Studies have shown that just looking at a meadow reduces heart rate and makes us feel happier.

In a regenerative farming system, meadow grasslands or herbal leys are a key part of providing important habitat plus nutritious forage for healthy animals, which in turn creates high-nutrient-density milk, meat and wool. Regenerative farmers Toby Diggens and Bella Lowes in Devon have innovatively used biodiversity net gain (BNG) payments to purchase a degraded former swede field and turn it into a species rich grassland and wildlife haven with scrapes to attract lapwing and plover, and thick hedgerows, all of which are browsed by their small herd of red ruby Devon cattle. The grazing, trampling and dunging of these herbivores has created a variety of open and dense sward and a complex mosaic of habitats is now filled with butterflies and insects.

There are as many different types of meadow as there are soils and microclimates, so the first thing to understand for your own is its context, which will dictate your best meadow archetype, from a herbal ley, wet meadow or chalk meadow. One of my favourite meadows is the medicinal meadow, which was often placed near the farmhouse. Such an area would be set aside for sick animals to come to graze, allowing them to choose the medicinal plants they needed for themselves. A medicinal meadow might include plants like wild angelica (*Angelica sylvestris*), bugle (*Ajuga reptans*), daisies (*Bellis perennis*), meadowsweet (*Filipendula ulmaria*), common stinging nettle (*Urtica dioica*), ribwort plantain (*Plantago lanceolata*) and germander speedwell (*Veronica chamaedrys*).

If you want to establish a meadow in an existing grassland it's a simple process, best done gradually over about four years. The first thing to do is to reduce the vigour of the existing grass, which will usually have a thick mat of roots difficult for seedlings to establish among. Scarify hard in strips and then sow yellow rattle (*Rhinanthus minor*) or red bartsia (*Odontites vernus*) seed or strew local donor hay with a high percentage of yellow rattle (or do both). (Note: red bartsia can be invasive in

Scarify hard in strips and then sow yellow rattle (*Rhinanthus minor*) or red bartsia (*Odontites vernus*) seed or strew local donor hay with a high percentage of yellow rattle.

Above: If you want to establish a meadow in an existing grassland it's a simple process, best done gradually over about four years.

Opposite: Studies have shown that just looking at a meadow reduces heart rate and makes us feel happier.

the USA.) If you don't have a local hay meadow that will give you some hay, use local seed merchants that have similar conditions to your own soil.

For a small or urban space, a quicker option is to buy a wildflower mat, which is lighter than turf and will give a good show in the first year. It will usually include some annuals like poppies (*Papaver*) and cornflowers (*Centaurea cyanus*), which may not come back in future years as more vigorous species like ox-eye daisies (*Leucanthemum vulgare*) and knapweed (*Centaurea nigra*) may dominate for a while, but they tend to find a balance over time.

The natural pattern of a meadow follows the contours and conditions of the land and usually include waves of species depending on sun, shade, moisture and wind. When planning one, add to this some vigorous species that create sporadic punctuation, and a lovely rhythm will flow.

PRAIRIE PLANTING

In both a naturalistic garden setting and in a city where basements below can create endemic compaction and buildings cause wind tunnels, the steppe or prairie archetype works well. The plants in such places experience hot summers and cold winters, with low humidity and tough soils. Steppes and prairies are similar, with the main difference being depth of topsoil and fertility, so select the best plants to suit your garden context.

These are tough plants used to minimal rainfall; they can go dormant when necessary, but will not tolerate extended wet conditions. Many prairie plants have thick mats of roots to intercept any rain.

Key emergent genera like feather grass (*Stipa*) or fescue (*Festuca*) grasses and key species like black-eyed Susan (*Rudbeckia fulgida*) anchor these plant communities, with deep fibrous roots that help to stabilise soil and provide a habitat for beneficial microbes.

Species like coneflower (*Echinacea pallida*) and rattlesnake-master (*Eryngium yuccifolium*) make good companions as they have deep taproots to access water in a layer below the others.

Pollinator plants with mat-forming roots like yellow pink (*Dianthus knappii*), yarrow (*Achillea*) and blanketflower (*Gaillardia*) or running roots like slender vervain (*Verbena rigida*) are all drought tolerant and attractive to beneficial insects as well as providing seed heads and structure to a planting scheme.

For ground cover, tough succulents like stonecrop (*Hylotelephium*) and *Delosperma* will protect the soil, while nitrogen-fixing legumes like rose clover (*Trifolium hirtum*) will enhance soil fertility.

In planning a prairie-inspired scheme, work with the colours of the surrounding landscape to make your planting harmonious. Structural plants like coneflower (*Echinacea* and *Rudbeckia*) typically grow in small clumps in the wild so dot them through large drifts of grasses like Lessing feather grass (*Stipa*

Structural plants like coneflowers typically grow in clumps in the wild.

Prairie plants' roots can rot in wet soil or self-seed extensively so apply a thick gravel mulch.

lessingiana) and aromatic prairie dropseed (*Sporobolus heterolepis*) with the occasional addition of the sea holly (*Eryngium*) and *Verbena* layer to make it feel naturalistic. The stonecrop (*Hylotelephium*) and nitrogen-fixing layer won't add much visually but is a vital forage layer for pollinators and for soil amelioration. Finally, bulbs like prairie onions (*Allium stellatum*) and red-hot pokers (*Kniphofia*) and summer-dormant oxlips (*Primula elatior*) will provide nectar pollen and colour in spring and tolerate being in the shade of other plants for the rest of the year.

In milder UK climates prairie plants' roots can rot in wet soil or self-seed extensively if too happy. A thick gravel mulch helps with both issues. The prairie archetype has few trees and is read as a wild open space, but in a garden setting smaller canopy trees or large shrubs like eastern redbud (*Cercis canadensis*) at the edges could add shade and create a transition to another area or planting type.

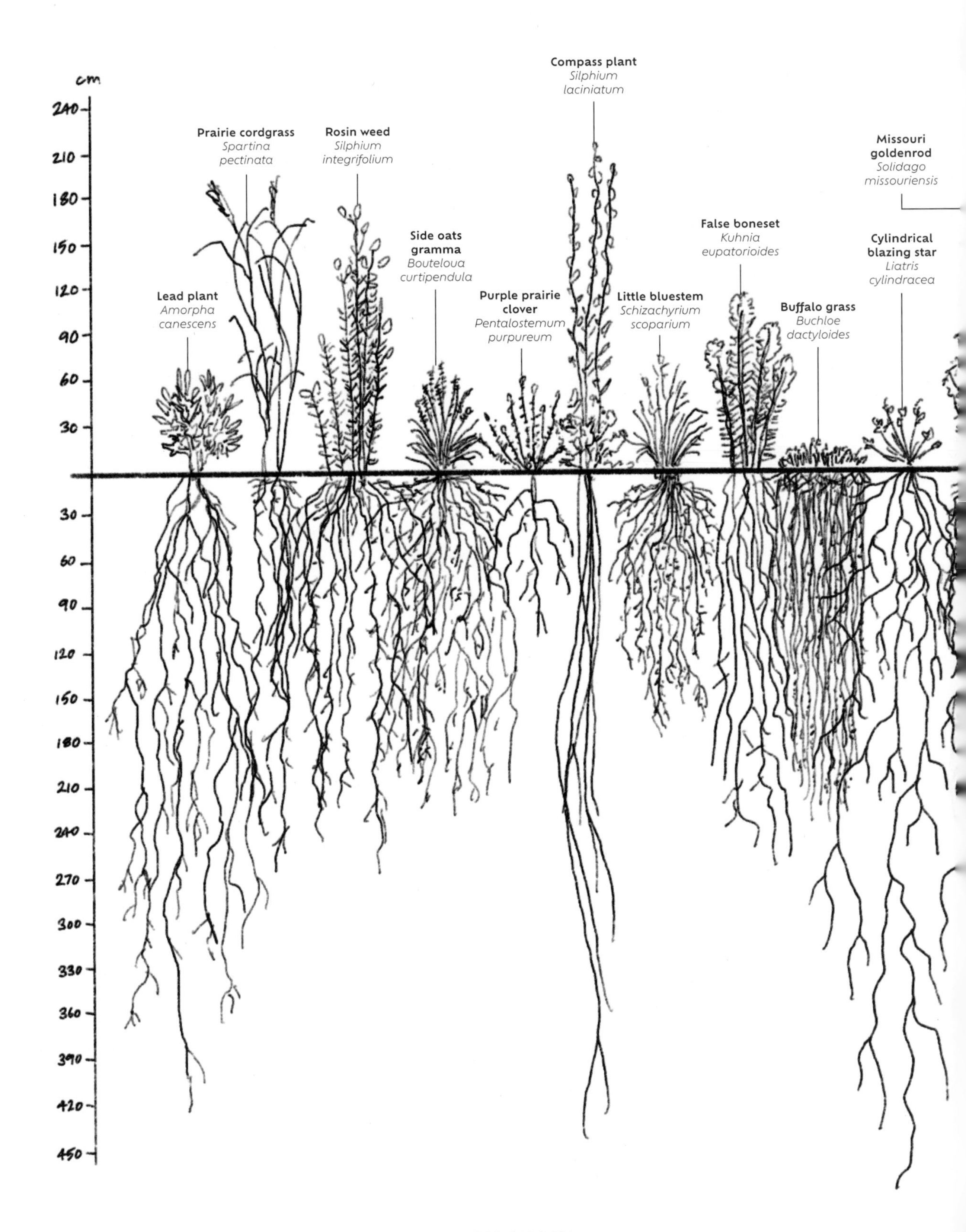
cm
240
210
180
150
120
90
60
30
30
60
90
120
150
180
210
240
270
300
330
360
390
420
450
Compass plant
Silphium laciniatum
Prairie cordgrass
Spartina pectinata
Rosin weed
Silphium integrifolium
Missouri goldenrod
Solidago missouriensis
Side oats gramma
Bouteloua curtipendula
False boneset
Kuhnia eupatorioides
Cylindrical blazing star
Liatris cylindracea
Lead plant
Amorpha canescens
Purple prairie clover
Pentalostemum purpureum
Little bluestem
Schizachyrium scoparium
Buffalo grass
Buchloe dactyloides

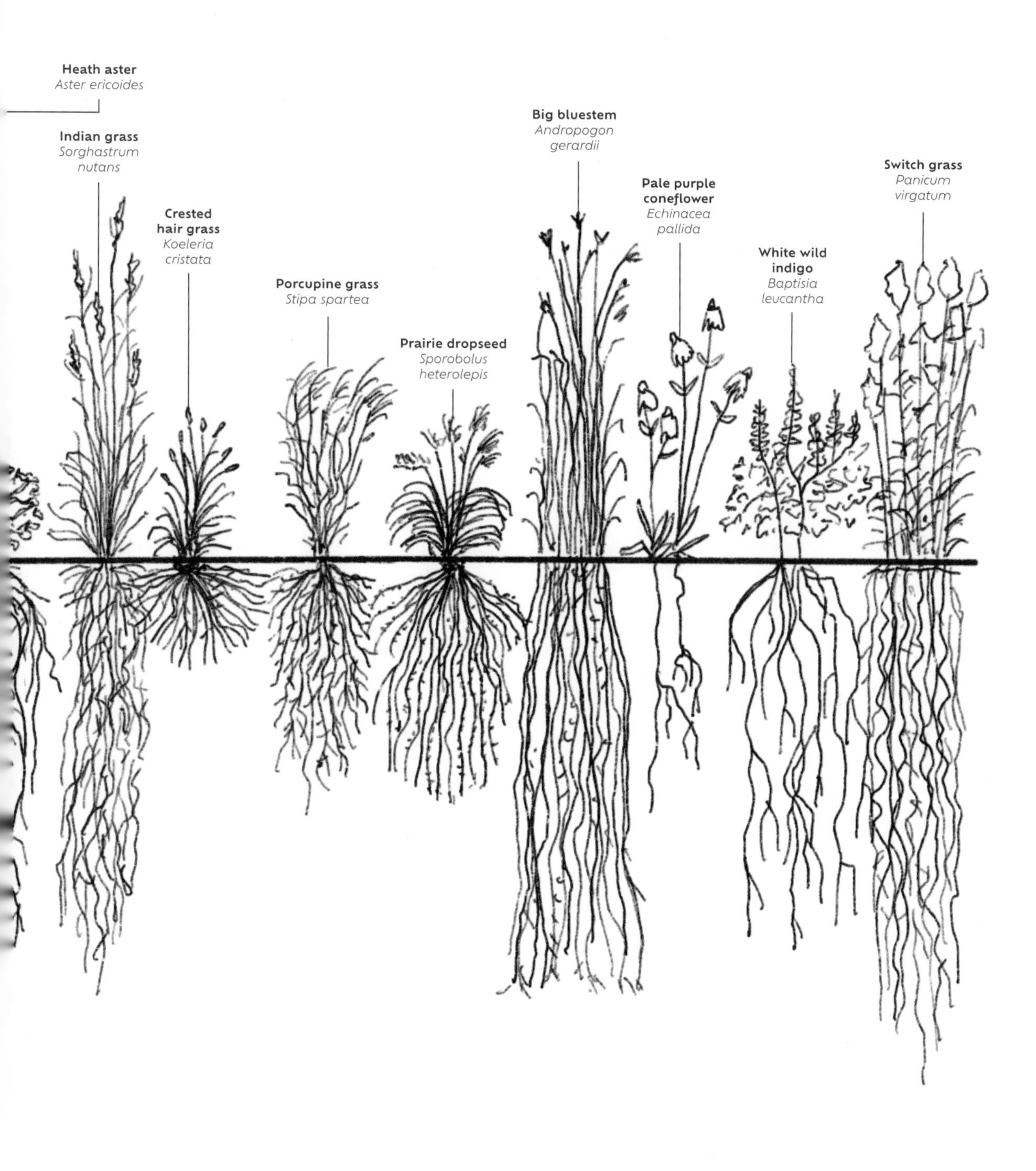

Typical Prairie Species.
Based on 'Root Systems of Prairie Plants', Heidi Natura – Conservation Research Institute, 1995

Above: These typically early succession plants will thrive on poor soil.

Opposite: Ideal sites are brownfield degraded soils, mounds of crushed builders' rubble, recycled aggregate or locally sourced gravel over existing soil.

MEDITERRANEAN PLANTING

Mediterranean plants have adapted to thrive in areas where soil is typically shallow and water drains easily. They tend to flower early to take advantage of rain, and can become summer dormant to endure drought. They may grow on rocky areas and be resilient to grazing by sheep or goats. Some are frost tender, but many will survive a colder winter as long as their roots do not sit in water. These preferences suit shallow-soiled city gardens with good drainage, where the heat-island effect keeps most winters mild.

Plants adapt to drought with strategies like waxy cuticles or fine leaf hairs to reduce transpiration (water loss) in the heat of the day. These plants tend to have roots that reach down to find water, or lateral roots to catch surface drops – and sometimes both. They grow low and slow in their harsh home climate so avoid high-nutrient soil, which along with too much rain will make them tall, floppy and short-lived.

These are typically early succession plants (happy in high bacteria, lower fungi soils) so will thrive on poor soil but will be out-competed by dense herbaceous planting and will rot and die in waterlogged clay. Plant in brownfield sites with degraded soils, or in mounds of crushed builders' rubble, recycled aggregate or a 20cm (8in) layer of locally sourced gravel spread over existing soil or overturned turf.

Choose small multi-stemmed trees such as snowy mespilus (*Amelanchier lamarckii*), star magnolia (*Magnolia stellata*) and crab apple (*Malus hupehensis*) to provide the structure you might find from olives (*Olea*) or slow-growing oaks (*Quercus*) in the Mediterranean. Choose a structural herbaceous layer to include evergreens like rosemary (*Salvia rosmarinus*), *Euphorbia* and silver-leaved lamb's ear (*Stachys byzantina*). Emergents like Scotch rose (*Rosa spinosissima*), *Calamagrostis* × *acutiflora* 'Karl Foerster' and great mullein (*Verbascum thapsus*) will provide height and attractive skeletons in winter.

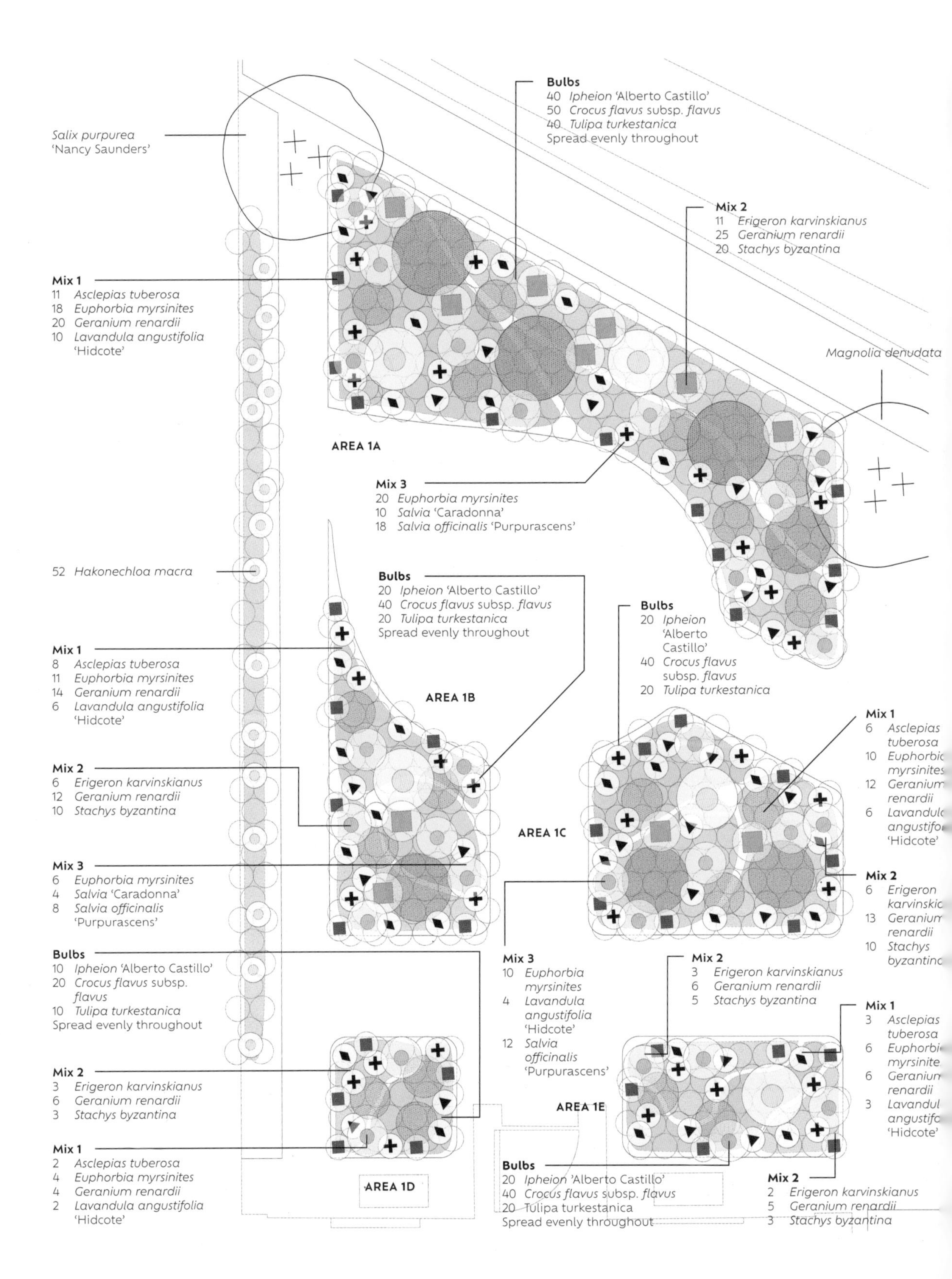

Bulbs
40 Ipheion 'Alberto Castillo'
50 Crocus flavus subsp. flavus
40 Tulipa turkestanica
Spread evenly throughout
Salix purpurea
'Nancy Saunders'
Mix 2
11 Erigeron karvinskianus
25 Geranium renardii
20 Stachys byzantina
Mix 1
11 Asclepias tuberosa
18 Euphorbia myrsinites
20 Geranium renardii
10 Lavandula angustifolia
'Hidcote'
Magnolia denudata
AREA 1A
Mix 3
20 Euphorbia myrsinites
10 Salvia 'Caradonna'
18 Salvia officinalis 'Purpurascens'
52 Hakonechloa macra
Bulbs
20 Ipheion 'Alberto Castillo'
40 Crocus flavus subsp. flavus
20 Tulipa turkestanica
Spread evenly throughout
Bulbs
20 Ipheion
'Alberto
Castillo'
40 Crocus flavus
subsp. flavus
20 Tulipa turkestanica
Mix 1
8 Asclepias tuberosa
11 Euphorbia myrsinites
14 Geranium renardii
6 Lavandula angustifolia
'Hidcote'
AREA 1B
Mix 1
6 Asclepias
tuberosa
10 Euphorbia
myrsinites
12 Geranium
renardii
6 Lavandula
angustifolia
'Hidcote'
Mix 2
6 Erigeron karvinskianus
12 Geranium renardii
10 Stachys byzantina
AREA 1C
Mix 3
6 Euphorbia myrsinites
4 Salvia 'Caradonna'
8 Salvia officinalis
'Purpurascens'
Mix 2
6 Erigeron
karvinskianus
13 Geranium
renardii
10 Stachys
byzantina
Bulbs
10 Ipheion 'Alberto Castillo'
20 Crocus flavus subsp.
flavus
10 Tulipa turkestanica
Spread evenly throughout
Mix 3
10 Euphorbia
myrsinites
4 Lavandula
angustifolia
'Hidcote'
12 Salvia
officinalis
'Purpurascens'
Mix 2
3 Erigeron karvinskianus
6 Geranium renardii
5 Stachys byzantina
Mix 1
3 Asclepias
tuberosa
6 Euphorbia
myrsinites
6 Geranium
renardii
3 Lavandula
angustifolia
'Hidcote'
Mix 2
3 Erigeron karvinskianus
6 Geranium renardii
3 Stachys byzantina
AREA 1E
Mix 1
2 Asclepias tuberosa
4 Euphorbia myrsinites
4 Geranium renardii
2 Lavandula angustifolia
'Hidcote'
AREA 1D
Bulbs
20 Ipheion 'Alberto Castillo'
40 Crocus flavus subsp. flavus
20 Tulipa turkestanica
Spread evenly throughout
Mix 2
2 Erigeron karvinskianus
5 Geranium renardii
3 Stachys byzantina

To create a naturalistic and harmonious flow, choose a planting mix and repeat it, rather than using a traditional planting plan of mono-species swathes. In the wild, plants tend to mix and find their own balance, so to mimic this add annuals like toothpick plant (*Visnaga daucoides*) threaded through with short-lived perennial golden columbine (*Aquilegia chrysantha*) and fennel (*Foeniculum vulgare*) to add the dynamism of appearing in different places as they seed about over the years. Bulbs provide forage and colour in spring: choose *Ipheion* 'Alberto Castillo', early bulbous iris (*Iris reticulata*) or Turkestan tulip (*Tulipa turkestanica*).

3 *Myrtus communis*

4 *Rosmarinus officinalis* Prostrata Group

5 *Potentilla fruticosa* 'Primrose Beauty'

8 *Euphorbia characias* subsp. *wulfenii*

13 *Dierama pulcherrimum*

17 *Pennisetum villosum*

9 *Verbascum thapsus*

37 *Thymus serphyllum* 'Pink Chintz'

33 *Nigella damascena*

21 *Eschscholzia californica*

30 *Allium christophii*

15 *Agastache rugosa f. albiflora* 'Liquorice White'

To create a naturalistic and harmonious flow, choose a planting mix and repeat it, rather than using a traditional planting plan of mono-species swathes. In the wild, plants tend to mix and find their own balance.

ESTABLISHMENT AND MANAGEMENT

> *The best way to take care of the future*
> *is to take care of the present moment.*
> *Thích Nhất Hạnh*

With more frequent extreme weather events, our planting will adjust naturally or need adjusting by us to suit our aesthetics. Less resilient plants will fade out or be overtaken by tougher ones. In extreme wet, a population spike in slugs and snails may decimate new shoots while a drought may cause new trees to drop leaves early to survive. Allow for this in the planning stage, building in flexibility by choosing several species or cultivars with similar functions rather than relying on just one. Also plan regular reviews to edit planting. Being flexible allows you to try new things, to observe and learn. You can loosen control and foster an adaptable mindset.

Below: Build in flexibility by choosing several species or cultivars with similar functions rather than relying on just one and plan regular reviews to edit planting.

Opposite: Source plants small. Younger trees will soon get going and overtake larger transplants.

Care at establishment sets plants up for success. Source plants small: younger trees will soon get going and overtake larger transplants. Buy plants in 9cm (3½in) pots for herbaceous plants or source bare root from specialist nurseries. Keep roots covered and moist between purchase and planting by wrapping or draping them in hessian or damp newspaper. Heel them into a bed of moist soil if planting is delayed by more than a few days. Only lay out the amount you can plant in a day to avoid plants drying out while waiting to be planted, and check roots as you go. Turn out each pot. If potting soil falls away from the roots, check for vine weevil infestations, and if roots are congested you can use a hook to pull them apart slightly and allow air in. When you plant, ensure each planting hole is not compacted; it should be loose on the bottom and sides to allow air and water to flow. Plant level with where each plant grew before; avoid burying it too deep, including into gravel. Add living compost and mycorrhizal fungus in the planting hole and companion plants above to create community, and mulch to protect the soil.

After establishment, management is about biomass and health. Planting in layers, incorporating native species, allowing self-seeder 'volunteers' and using a variety of substrates from deep organic matter to crushed aggregate will mitigate against extreme weather.

After planting, avoid air pockets in the soil, which will cause root dieback: gently press soil in with your foot around the base, and ensure a good drenching to help soil to move into any cavities.

Sowing locally sourced, open-pollinated seeds, either your own or from similar gardens or specialist nurseries, is a wonderful way to introduce diverse designs and to select species that are thriving and adapting to your conditions. Soak them in living compost or drip humic acid or compost tea (see page 63) into planting beds as you sow to give them a flying start. This is a good way to mix annuals and ephemerals into a dynamic new scheme.

Although smaller plants mean less water is needed during establishment, it is still vital that young plants don't dry out for the first season. They require water to get their roots extended and the all-important rhizosheath established, which will buffer them from fluctuations in moisture, pH and minerals as they grow. Give a good drenching once a week rather than frequent sprinklings to allow water to penetrate and encourage deeper roots. In dry areas form a trench or bowl in the soil around new planting to fill with water, and allow it to gently percolate. In very wet areas, plant trees and shrubs on slight mounds to keep their crowns out of waterlogged soil.

After establishment, management is about biomass and health. You should aim to accumulate biomass to store carbon, water and nutrients. This can be as simple as by cutting plants and leaving them to decompose *in situ*, or by composting them to add more plant-available nutrients. When pruning we are being human herbivores: taking an occasional nibble of lots of areas will allow gentle regrowth while a drastic hack will result in a big reaction.

Root depths, functional diversity, local adaptation, ecological relationships and mimicking natural associations all contribute to a planting scheme's overall resilience. Planting in layers, incorporating native species, allowing self-seeder 'volunteers' and using a variety of substrates from deep organic matter to crushed aggregate will mitigate against extreme weather. You can hedge your bets against what the future might bring, while still enjoying observing what is happening every day.

Compost, wool and woodchip are all valid mulches depending on your context.

MULCH

Once we have understood the importance of looking after soil, the need for ground cover or mulch becomes obvious. Mulch protects soil from compacting rain and run-off, retains moisture, suppresses weeds, insulates from cold in winter and keeps roots cool in summer. Aesthetically a layer of composted wood mulch makes a border look finished and tidy too. I follow the maxim that if you mulch the first 30cm (12in) from the edge all 'weeds' further back will be forgiven.

Having decided to mulch, the question is with what. Compost, wool and woodchip are all valid choices depending on your context. These all provide nutrients to the soil and plants as they break down. If you have living compost (see page 56) this provides a full diversity of microbes and will help the soil food web establish in your borders. It can be spread around plants in a thin layer or be sprinkled over ground cover to fill in gaps. Wool is a natural product from sheep and easy to buy or may come free in packaging. It contains some nitrogen, which will be released gradually to feed plants as it breaks down, as well as plenty of carbon, which will improve soil structure, water retention and nutrient availability. Since wool has near neutral pH it can buffer soil fluctuations too, and its micronutrients include sulphur, zinc, iron and copper, which are all essential for plant health.

For plants at the stage of succession that will benefit from higher fungal to bacterial ratio (see page 48), like shrubs and trees, woodchip is ideal. It has the same protective properties as compost and wool, plus the additional benefit of being high in lignin. Lignin is vital to humification; as it breaks down it is key to the process of creating living soil or humus. When choosing woodchip always ask which trees it is from. Conifers like pine (*Pinus*), cedar (*Cedrus*), coastal redwoods (*Sequoia sempervirens*) and cypress (*Cupressus*) possess a resin that is chosen for its insect-repelling properties around houses and a slow decomposition time – neither of which is a helpful asset in regenerative soils. Likewise, gums (*Eucalyptus*) have allelopathic properties, so can actively slow down growth of other species – an asset only for paths.

For new hedging, deciduous woodchip is a game changer in the establishment phase, suppressing weeds and reducing requirement for watering. For woodland-edge planting and orchards, ramial woodchip from deciduous broad-leaved trees is ideal. Ramial chipped wood is from the smaller branches of a tree, which hold the most nutrients (nitrogen, phosphorus, potassium, calcium and magnesium). If you are chipping it yourself, choose branches about 7cm (3in) in girth. Branches without leaves are easier for a small chipper and will be slower to decompose, having less nitrogen.

If you are lucky enough to have willow (*Salix*) or can grow some, enjoy the medicinal qualities of its woodchip. There is mounting evidence that the salicin in willow helps disease resistance in other plants and can prevent scab caused by fungus in apples. As the willow breaks down, the salicin is taken up by tree roots. It also helps combat early blight in potatoes, powdery mildew, fire blight and fungal pathogens like phytophthora. A plant that grows easily in heavy wet soils, willow can be used for weaving baskets, fences and play domes. It is also a great food source for early bees, caterpillars and the endearing willow warblers and willow tits.

Willow is a great mulch to grow both for your soil and also for wildlife like the willow warbler.

PART TWO

MEASURING OUR IMPACT

WHY SHOULD WE MEASURE OUR IMPACT?

The important thing is not to stop questioning. Curiosity has its own reason for existing.
Albert Einstein

The push to get to net-zero carbon emissions and to slow biodiversity loss has led to an industry of carbon measurement and offsetting, biodiversity net-gain credits and water-neutrality credits. At worst these can be traded to allow individual harm still to be done in the name of the collective good. At best they are excellent decision-making tools to use when making choices – our allies in knowing whether what we are planning is healing or harmful, and to what extent we are succeeding in regenerating the land.

So how do you know if you are doing good in your own garden and designs? You might know it in your gut, through intuition, and through the physical transformation you see and feel. There is also a place for facts and figures to answer to the head as well as the heart.

Whether for the largest regenerative project or the smallest garden, the first thing is to understand what is there already – your baseline. This is an enjoyable process of getting to know your land and all who share it with you, to log how healthy it is physically and energetically, and where there may be issues that need help.

Start with a topographical survey that maps the size of the land, the above- and below-ground services, and the hidden geology, waterways and soils. Look at historic maps and geological websites, plus sites like the Multi-Agency Geographic Information for the Countryside (MAGIC) maps and Land App. An arboriculture survey will document the health of trees

Above: An arboriculture survey will document the health of trees and root protection areas for any planned construction. Ecological surveys map who else lives where; soil tests explain available minerals and biology.

Opposite: Tests for rivers, lakes and streams detail water quality and nutrients.

and root protection areas for any planned construction. Ecological surveys map who else lives where; soil tests explain available minerals and biology; and tests for rivers, lakes and streams detail water quality and nutrients. Sap tests show how well vegetables and crops are photosynthesising, and whether their growing environment could be improved.

Once you have a baseline, plan reviews as part of your management plan. See what impact your approach is having, and how it could develop. Whatever the size of garden, it's edifying if we get it wrong and hugely rewarding when life and health floods back in.

When planning the planting, measure how much mains water is used in the house and how much you have left to water the garden if you are to stick to the UN's recommendations. Assess how much you use, and how much rainwater you can collect and store; then plan planting to suit. When buying furniture or planning hard landscaping, assess the environmental footprint of the materials you use by referring to EPDs (see page 127), and use the tables on pages 128–9, 135 and 228 as a guide.

In 2024 biodiversity net gain became part of UK planning law. This means that every planning application larger than a householder application must show that it is providing a 10 per cent, or more in some areas, increase in habitat than prior to development, secured for thirty years. Whether to satisfy planning requirements or just your own interest, it is instructive to measure how much good or harm you are doing. In my studio we have developed a simple tool for gardens to measure the impact of new interventions, based on how much habitat was there before and how much you can add.

There are also several simple measurements for soil health you can do on site, which we will look at next.

WHAT CAN WE MEASURE IN OUR OWN GARDENS?

Count the number of types of soil life.

> *The measure of intelligence*
> *is the ability to change.*
> *Albert Einstein*

Tests and evaluations take a snapshot in time to help us make decisions. It is not about being good or bad, but showing a pattern, allowing us to decide if we should do something to intervene, manage totally differently or do nothing – an equally viable solution.

In your garden choose an area that is thriving and one that is doing less well and test them both through the seasons to compare management. Here are some things to measure so you can understand what they are telling you.

1. Diversity of plant life

Mark out a square with sides of 1m (3ft) (or one of any size as long as you are consistent). Count the number of different ground-cover plants in the area. Estimate how much bare soil there is. Make notes and compare through the seasons.

- Ground cover protects the soil, and the more diversity the more resilience there is.

2. Diversity of soil life

In the same square, count the number of types of soil life – look carefully under leaves and litter. (Many millipedes or many spiders count as one type.)

- The variety of life shows how healthy your ecosystem is.
- You might note that you have ground-nesting ants and spiders in bare ground, indicating that some bare soil can be good too.

3. Soil pH

Use a soil pH test kit available online or from garden centres. Take a small amount of soil and add it to the test tube along with the solution – or as per instructions on the pack. Check the colour against the chart to see how acid or alkaline your soil is. This will depend on your geology and your soil's stage of succession (see pages 48–9). Check a pH nutrient chart: if your soil is not neutral (pH7), add living compost to get the microbes cycling.

- Soil pH affects which nutrients are available to your plants. Many nutrients become less available if pH goes above 7.5 or below 6.
- Later succession plants, which prefer a fungal-rich, old, acidic soil, will not thrive on a newer or chalkier soil, and vice versa.

4. Water infiltration

Cut a length of 15cm (6in) diameter polyvinyl chloride (PVC) or steel pipe to 15cm (6in) length. If using PVC you may need to bevel the edges to make them sharper. Allowing forty-eight hours after any heavy rain, place the ring

An infiltration test checks for compaction by measuring how fast water is able to infiltrate.

on your lawn and cut away vegetation under the ring to allow it to dig in. With a flat piece of wood on top of the ring, use a mallet to bang the pipe evenly into the ground to a depth of 2cm (¾in). Gently pour 500ml (17½fl oz) of water into the centre and set the timer to see how long it takes to disappear. Once it has all gone, repeat the test with a second 500ml (17½ fl oz) of water and note the time again.

- Repeating this in different areas and through the seasons shows how easy it is for water to enter and stay in your soil. A healthy soil with plenty of humus will hold water available for plants, acting like a sponge.

5. Compaction

Use a professional penetrometer or make one out of a 50cm (20in) length of 3.15mm (10-gauge), high-tensile wire. Use 12cm (5in) of the length to make a curving handle and on the remaining 38cm (15in) make marks every 2.5cm (1in) from the end. Push the end of the penetrometer slowly into the ground using firm but gentle pressure. Record in cm (in) the depth you can penetrate with even force. Try this on a lawn, in a border and a grass path to see the difference. Start again if you hit a tree root or stone.

- The further you can penetrate, the easier it is for air, water and soil life to flow and for roots to find grow and source nutrients. Once you have completed step 6 (below), test the penetrometer again, starting at the bottom of the 20cm (8in) hole. How much farther or less far can you go than before? Is it harder to get through the subsoil, or have you broken through a compaction pan to allow deeper penetration? Note what this would mean for the roots in this area.

6. Rooting depth and rhizosheaths

Cut a cube, 20×20×20cm (8×8×8in), of soil and dig it out in one block. Try to keep it together and place it on a piece of hessian or an old bag. See how deep the roots of any plants at the edges of the cube go and whether they are going down straight or creeping sideways to avoid a layer of compaction. Pull the cube apart and hold the plants up to see if the roots are coated in humus to form a healthy rhizosheath. Are they single strands or do they have some aggregates (clumps of soil) clinging to them? Or do they look like full dreadlocks with lots of healthy buffering humus?

- Deeper roots can access more minerals, nutrients and water from bedrock.
- A healthy rhizosheath will protect and buffer roots from changes in pH, drought and attacks by insects or toxins.

A healthy rhizosheath will protect and buffer roots from changes in pH, drought and attacks by insects or toxins.

Count the number of types of soil life to see how healthy your soil is.

7. Soil structure

Dig a small handful from the top 5cm (2in) of soil beside the hole dug for the cube and another 5cm (2in) from the bottom of the hole. Notice the size and arrangement of aggregates or clumps. Arrange the clumps loosely in size and note what size is predominant; what is the biggest and smallest clump; do they break into crumbs of about 1cm (½in) if you push down on them with a firm finger? Or does your finger make an indent in a layer of solid clay? Perhaps it falls apart like sand? Compare the cube to other areas of the site and take photos for next season.

- Poor soil structure means clumps being too solid or too loose.
- Hard crumbs, solid layers or big clods point to compacted soil, which is difficult for air, water and nutrients to move through.
- If soil is too loose it can't hold on to nutrients and water.
- The solution to both problems is to add living compost (see page 56), and to introduce tough plants so their roots can penetrate and aggregate the soil and create humus.

8. Slake test or aggregate stability

Fill a lidded jar with 125ml (4 fl oz) of water. Drop in three or four pea-sized soil aggregates and allow to stand for 1 minute. Note if they fall apart or stay together. After a minute gently swirl the jar and see if the aggregates still stay intact. Repeat the swirling movement more vigorously if they continue to stay together. Compare different areas of your garden, and note at which point the aggregates disintegrate.

- Healthy soils usually have more stable aggregates, held together with plant exudates and fungi. As your soil becomes more healthy, notice the difference.

9. Earthworm and insect count

Using the 20cm cube (see 6 above) count the number of earthworms and any other species of insects and arthropods you find.

- Earthworms are an indication of healthy soil as they thrive in neutral-pH, humus-rich, moist soil with low chemical residues (also good plant conditions). Such worms help the soil by feeding on organic matter and processing it into plant-nutritious vermicast.
- Note other insects and arthropods you find; look up whether or not they are beneficial to your plants and which predators you may need to call in.

A brix test tells you how well your plant is photosynthesising and whether it needs help.

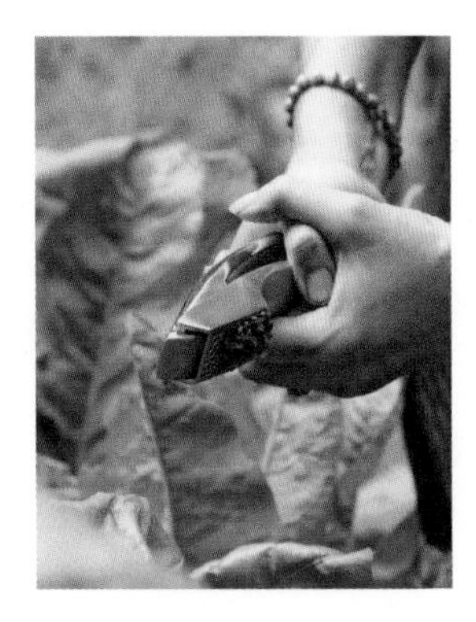

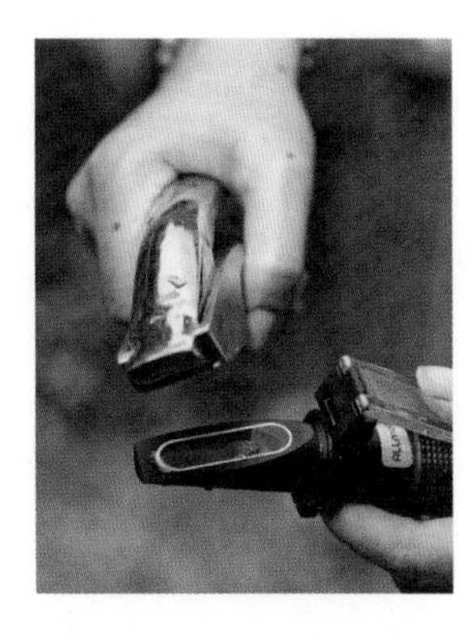

10. Brix

Essentially brix tells you how well your plant is photosynthesising and whether it needs help. To measure brix – that is, the amount of sugars in a plant – you need a simple refractometer. First clean the refractometer plate with distilled water. Then use a garlic crusher to squeeze a drop of sap out of your chosen leaf or berry. Drop the liquid straight on to the refractometer plate, then close the lid and allow the liquid to spread. Hold the refractometer up to sunlight and read the measurement (where the blue gets up to a certain line.)

- If brix is low and you are growing commercially, you may want to send leaves off for sap testing to see which minerals they are lacking and investigate why.
- It can help to measure your favoured plant and a weed nearby to see which is photosynthesising better. Do the conditions favour the weed or your plant?

11. Soil biology and minerals

Send your soil off to a lab to give you a breakdown of the soil biology.

- Once you know where your soil is on the succession chart (see pages 48–9) you can plant to suit your current soil health.
- Send your compost for analysis too and begin to add beneficial organisms through living compost inoculants, protist teas and additions like fish hydrolysate or lactobacillis (LAB) – see page 63.

HOW TO ASSESS YOUR IMPACT

Gardens can be very complex despite being small in scale, and can offer a great contribution to local biodiversity.

> *Don't guess what you can measure.*
> *John Kempf*

In large-scale food growing, yields and costs are meticulously measured both to balance the books and to learn from the past to improve the future. One of the dangers of the 'bigger is better' mindset, however, is that it can lead to a one-size-fits-all mentality. Our gardens and landscapes are so diverse that the context changes between watersheds and counties, even over a hedgerow. Rather than creating a big dataset to roll out at scale, we can measure what is working for us, and what we can learn and share. This holds true for large-scale regenerative projects and smaller gardens.

The idea in simple form is to document the wildlife that exists in terms of flora and fauna on a piece of land and then to compare the increase that is achieved by adding appropriate planting, habitat, water and connectivity. In the case of the new BNG regulations (see previous section) the landowner commits to managing the improvement for thirty years in order to receive payment for this public good.

Gardens can be very complex despite being smaller in scale, and can offer a great contribution to local biodiversity. The UK Government biodiversity net gain calculator is not yet geared to residential gardens, so for these we can measure our impact using a simple biodiversity calculator like the one below.

Biodiversity gain calculator

Use this calculator to plan and measure how much impact you have already created and how much more could come.

Once you have seen the positive impact that complexity in planting and biodiversity has on weather resilience and our own health and feelings, very few creatures seem like pests any more; nor are any plants weeds. However, you can use the chart opposite to help decide how to manage any over-vigorous volunteer plants.

BIODIVERSITY GAIN CALCULATOR

Biodiversity Score: 0–5
0 = lowest biodiversity value; 5 = highest biodiversity value

Step 1: Fill in the area (in square metres) for each planting type currently in your garden.
Step 2: Multiply the Biodiversity Score by the area for each planting type and write the value in the Total Biodiversity Score column.
Step 3: Add up the Total Biodiversity Score column to give the total score for the garden.
Step 4: Repeat steps 1 to 3 for your proposed garden design.
Step 5: Subtract the 'Before' garden score from the 'After' garden score, and divide this value by the 'Before' garden score.
Step 6: Multiply this value by 100 to give the Biodiversity Gain as a percentage – aim for as large as possible!

BEFORE: YOUR GARDEN NOW

Planting Type	Biodiversity Score	Area (sqm)	Total Biodiversity Score
Mown lawn	1		
Non-native hedge	1		
Vegetables	2		
Green roof (e.g. sedum)	2		
Non-native tree (non-flowering)*	2		
Gravel garden	3		
Non-native flowering tree*	3		
Young native tree*	3		
Wildflower meadow	4		
Native or edible hedge	4		
Native scrub	4		
Biodiverse planted roof (e.g. wild flowers)	4		
Edible perennials or herbs	5		
Wetland	5		
Water (ponds or streams)	5		
Veteran native tree*	5		
		Current Garden Score	

For trees measure approximate canopy area

AFTER: YOUR FUTURE GARDEN

Planting Type	Biodiversity Score	Area (sqm)	Total Biodiversity Score
Mown lawn	1		
Non-native hedge	1		
Vegetables	2		
Green roof (e.g. sedum)	2		
Non-native tree (non-flowering)*	2		
Gravel garden	3		
Non-native flowering tree*	3		
Young native tree*	3		
Wildflower meadow	4		
Native or edible hedge	4		
Native scrub	4		
Biodiverse planted roof (e.g. wild flowers)	4		
Edible perennials or herbs	5		
Wetland	5		
Water (ponds or streams)	5		
Veteran native tree*	5		
		New Garden Score	

For trees measure approximate canopy area.

Biodiversity Gain
Aim for a positive % change – as large as possible! If the percentage change is negative, the proposed garden design is worse for biodiversity than the existing garden. Remember, this is a planting type guide only and there are other features that can be added to enhance biodiversity, from bird boxes to log piles.

PART THREE

CONNECTION
THE KINDEST GARDEN

OUR NEED FOR CONNECTION

> *The world is full of magical things patiently waiting for our senses to grow sharper.*
> W.B. Yeats

Have you ever sat beneath an oak (*Quercus*) tree and wondered what it's thinking? What has it experienced during the few hundred years that it has stayed on that spot, with about 200 other species living within it? What challenges has it faced and how has it reacted? What has it processed and done?

Tree time is probably not the same as human time. Plants don't move around, and we humans associate moving with doing, but studies have shown that the electrical impulses used by plants to communicate are the same as those used in our own neuroreceptors. Plants don't have brains, neurones or synapses as we do, but they do respond to stimuli like light and insects; communicate with soil microbes, each other and insects; choose strategies to survive like early leaf drop; and learn. A speeded-up film of growing plants shows that what appears to us as stillness is in fact a frenzy of activity, but in a different register, leading Professor Jack C. Schultz to conclude that 'plants are just very slow animals'.

In the 1990s Dr Suzanne Simard established the concept of the mycelial web, now accepted as the way trees 'communicate above and below ground'. In 2018 Dr Monica Gagliano's study on pea plants showed how they reacted to the sound of running water, and would grow their roots towards the sound of real water but not towards the sound of a recording.

Heliotropic plants like sunflowers (*Helianthus annuus*) turn their heads to follow the sun, but the smaller tree mallow

Listen carefully. Can you perceive the century-old wisdom of the oak (*Quercus*), or understand the register of the mayfly, whose time frames are so much faster than ours?

(*Lavatera cretica*) will follow the sun during the day and then get into position to face the next sunrise during darkness, remembering where this moving target will appear even before the sun is up. In 2022 Dr Paco Calvo showed how the sensitive plant (*Mimosa pudica*) and carnivorous Venus flytrap (*Dionaea muscipula*) react to anaesthetic like a human or dog: when their senses are numbed, they no longer perceive the stimulus of touch. Humans make anaesthetics from plants, and plants also manufacture their own; trees exude the anaesthetic ethylene when their roots are cut, for example. Plant neurobiologists are showing how we can work with these discoveries to connect with plants, learn from them and live better with them. As Dr Paco Calvo said, 'to speak about plant intelligence is not taboo any more'.

Through history we have connected to the wisdom available through plant medicine, and in recent years there has been a revival as people seek connection through psychedelic plants such as psilocybin or the psychoactive brew ayahuasca. Many people are re-evaluating personal measures of success to something beyond social status or economic value. It can take a personal tragedy or setback to catalyse our understanding of connectedness to each other and the natural world. When financier Ben Goldsmith's daughter tragically died, he took a deep inner journey and with the help of plant medicine saw that we are all connected. He committed himself to nature, starting with his own estate in Somerset and expanding to invest in and promote large-scale regenerative projects worldwide through his podcast *Rewilding the World with Ben Goldsmith* and his book *God is an Octopus: Loss, Love and Calling to Nature*. Ben's vision is to allow beavers and cattle to roam across Selwood Forest as ecosystem engineers. His tiny herd of longhorn cattle have 'no fence' collars, which track them on an app. His land is an enchanted valley of beaver-linked ponds, swooping swallows and goldfinches feasting on thistle down.

Mycologists Paul Stamets and Merlin Sheldrake have explored the symbioses between fungi and humans, the way that we are interconnected, and how fungi may have shaped our existence in ways we don't understand. Paul and Merlin have published the mental and scientific breakthroughs that they have experienced when working with psilocybin mushrooms, a plant medicine that recent studies show can also help soldiers suffering from post-traumatic stress disorder (PTSD). Although many theorists attribute the development of early human cognitive development to our use of tools or cooked food, others including Terence McKenna propose that humans made the leap from apes to greater consciousness by ingesting psilocybin. Stamets and Sheldrake even suggest that the recent increase in discoveries of new psilocybin species may be linked to the collective leap of consciousness now needed to reconnect us with nature.

Without exploring plant journeying, most 'normal' medicines are based on plants. By understanding their healing properties, we can direct our eating to be therapeutic, just as cows will browse on willow (*Salix*) or meadowsweet (*Filipendula ulmaria*) for its healing salicylic acid, the base compound of aspirin, and dogs and hens will eat cleavers (*Galium aparine*), a zinc-accumulator plant. We are familiar with the effects of tea and coffee, which stimulate the brain, and plants like cacao (*Theobroma cacao*), which have been used for millennia in heart opening and healing ceremonies, and more recently for its comforting properties in sugary chocolate bars. If you have ever eaten a whole packet of biscuits, you will know the dopamine hit that wheat gives; wheat is a plant that has become so amended, addictive and indispensable as a food source that it is the second most traded grain in the world.

To connect with your land, see what it is offering each season. Try lime (*Tilia*) tree leaves in salads; lime tree flowers or elder (*Sambucus nigra*) flowers or berry syrup for a cold; rosehip syrup for vitamin C; greater burdock (*Arctium lappa*) root as a cleansing diuretic; common stinging nettle (*Urtica dioica*) leaves as a source of iron and cleansing, with seeds full of protein; lemon balm leaves (*Melissa officinalis*) for anxiety; mint leaves (*Mentha*) for digestion; and Midland hawthorn (*Crataegus laevigata*) flowers for a broken heart. So many plants are edible, and eating them connects us and our gut to the land.

Plants also connect with us. The most obvious is via seed. Cleavers and burdock seeds hitch a lift by means of tiny hooks, and tomato (*Solanum lycopersicum*) seeds thrive best after being through our gut. Anthropologists have shown how trees migrate to more favourable climates using birds and humans to transport seeds, and some plants can even manipulate other species. The cordyceps mushroom is a grim example. Its spores penetrate an ant's exoskeleton and compel the ant to climb to the top of a plant, clamp down on a leaf and wait for the mushroom to grow through its brain and rain down on the next generation of ants below.

Beyond eating them, being in company with plants is good for us too. Forest bathing (*shinrin-yoku*) is now prescribed in Japan and has been scientifically proven to reduce asthma, heart-rate issues and stress. Depression and anxiety can be alleviated by spending time in nature, and nature meditation is proven to be good for mind and body. Bringing plants inside homes and offices lifts the mood of city dwellers, and forest schools for children are having a resurgence as parents and teachers notice improvements in behaviour, alleviation in symptoms of attention deficit hyperactivity disorder (ADHD) and improved learning capacity. Doctors can prescribe gardening for mental

Make a mandala from petals and leaves, or keep a garden diary with drawings, pressed flowers and notes of what is happening each month.

health and weight loss, and fresh air improves sleep. About 20 minutes of sunlight boosts vitamin D, which is vital for immunity, and post-surgery healing is faster when we spend time in nature or just look at trees and plants.

We may smile when a new scientific study proves what gardeners have long known through experience and anecdote to be the case, but it is useful to have this logical endorsement – particularly for left-brain thinkers. Technology and scientific advice alone are not enough, though, to reconnect us and change the way we view and treat the world collectively. By developing our right-brain thinking, we can regain the intuitive connection with nature that our ancestors had.

As well as logic, or logos, we can weave in mythos – that way of exploring eternal truths through stories and metaphor, poetry, songs and art – to connect to the something else that many of us feel in the presence of a beautiful sunset, a huge wave breaking on the shore or an intricate pattern of a flower.

Listen closely. Can you perceive the centuries-old wisdom of the oak (*Quercus*) tree, or understand the register of the mayfly, caddisfly or male ant, whose time frames are so much faster than ours that in their short adult phase they have no need for working mouths? Sit with water, and fathom how it holds energy. Watch microbes in compost under a microscope. Tap into the sounds of trees and plants – with technology, their chemical communications become something that we can hear and even play music with.

Make a mandala from petals and leaves, or keep a garden diary with drawings, pressed flowers and notes of what is happening each month. Or study the lives of our smaller garden dwellers. For example, have you ever watched a carder bee 'carding' a lambs' ear (*Stachys byzantina*) leaf? They comb tiny fibres out of the soft leaves to weave their nests. Or watched slugs making love? It's very slow and very tender. Once we begin to empathise with our fellow Earth-dwellers, we can soon find that magical connection we need in all things.

HOW GARDENING FOR NATURE CAN BE GARDENING FOR OURSELVES

> *And I am thinking: maybe just looking and listening is the real work.*
> *Mary Oliver*

We are the human element in nature. Like plants we need enough air, water, sunlight and good soil to thrive. As growers we notice the effect of the Earth's yearly cycle around the sun, on temperature and day length each month, its impact on plants and even on our mood.

The moon cycles are equally influential. The moon's thirteen annual cycles of twenty-eight days around the Earth have been used by civilisations as far back as the Mayans to plan their crops. The gravitational pull of the moon not only creates tides that vary in size depending on how close it is to Earth; our fertility, energy and sleep patterns also follow these cycles, and many myths, moon goddesses and ancient magic reflect this.

In a garden context we can follow the moon cycles to help plant growth. Plant seeds and shoots in the first quarter of the cycle, when the moon is waxing and pulling plants up gravitationally, and plant roots and prune in the last quarter, when the moon is waning. At full moon rest and give thanks, like the Sri Lankans, who celebrate every full moon. The biodynamic system pioneered by Rudolph Steiner follows these principles, as per the chart overleaf. Once tried, the results are compelling. You may feel too busy to embark on this method, but the regular rhythm is easy to follow. Knowing what is happening on the cosmic scale brings us a little closer to understanding the land too, and the symbiosis between plants, ourselves and wider forces that science doesn't yet understand. Trusting in nature's cycles also helps keep us grounded among a barrage of fear-inducing news in the media.

Scientists have proven that gardening is good for us. It keeps us physically fit, burns excess calories and keeps our muscles moving and breath flowing. Just 20 minutes of raking has been shown to improve self-esteem. The joy of seeing new seedlings emerging or a butterfly drinking nectar from a flower is hard to beat. These all produce the feel-good hormone dopamine, while touching soil releases serotonin, which reduces stress.

Trusting in nature's cycles helps keep us grounded.

The corresponding effect of our touch on the soil has not yet been proven, but it is becoming clearer that there is no isolated 'human health'. We are interdependent. Our bodies are a microcosm of the wider biome and self-organising, intelligent communities that live around and within us. Connection to these communities is as essential to our health and energy as connection to our own kind.

So how can we connect? The first step is to take notice. Acknowledge, communicate and cooperate with the other beings around you. By communicate I don't mean verbal speech – it's more an energetic listening, like the listening that one is forced to do when you've driven yourself so hard that you collapse with stress and illness. Long before those extremes happen,

Follow the Moon cycles to help plant growth. Plant seeds and shoots in the first quarter of the cycle, when the Moon is waxing and pulling plants up gravitationally, and plant roots and prune in the last quarter, when the Moon is waning. At full moon rest and give thanks.

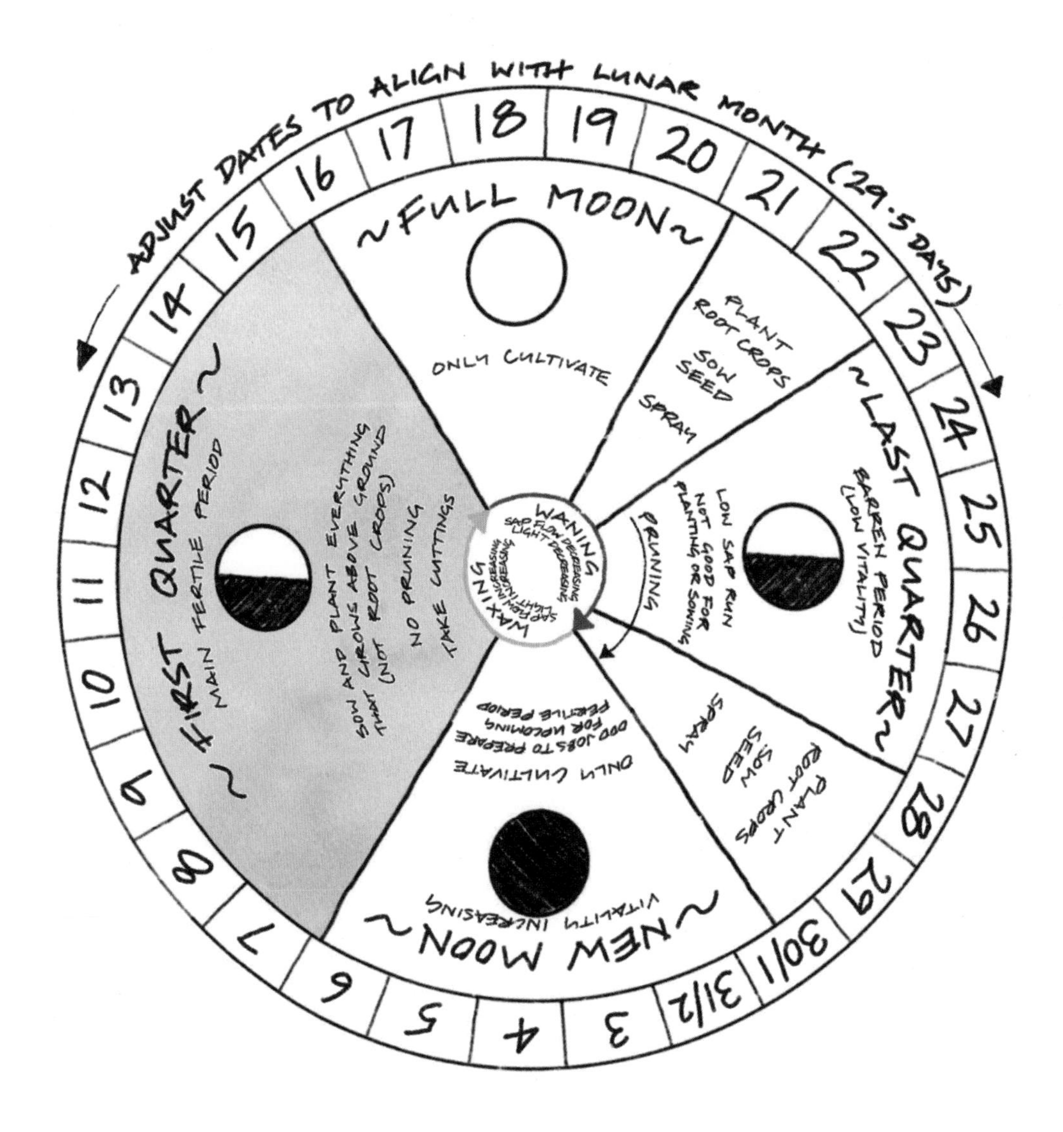

there are signals telling you that you are overdoing it, that life is becoming less joyful, that it's a struggle when it need not be. These signs indicate we should take time to look at what we are feeding our minds and bodies through food, air, light, rest and companionship, and the same is true with our plants and gardens.

Through a garden we can reconnect with the inherent wisdom of our body's natural community, the organs and organisms and microbes that live within us just like the microbes in the soil, and that need to be looked after just like the soil. The link between our own gut microbiome and that of the soil is becoming clearer as scientist pioneers like David Montgomery, Anne Biklé and Giulia Enders have explored how our physical and mental health are affected by what goes into our food. We trade microorganisms through absorption, inhalation and ingestion, through food and excretion. Once in our intestines, these microbes feed the human gut microbiome, which keeps the rest of our organs well. By growing healthy plants that photosynthesise to the best of their capabilities, we can produce nutrient-dense food. This means that it's full of health-giving secondary metabolites – the natural chemicals and minerals that give plants their colour, taste and smell, which draw us to them because they are good for us.

Studies have shown that a single spinach leaf contains over 800 different species of bacteria that it gets from the soil and the environment. By increasing the microbes on the leaves, the plant is protecting itself and making itself healthier, and when we eat the plant we are adding them to our own gut biome and increasing our own health.

Here are a few ways gardening regeneratively can benefit us. Add to them as you discover more:

- stress reduction – follow a rhythm and don't worry about results;
- improved mental health – connection, joy, sense of trust and proportion;
- better sleep – fresh air, exercise, listen to circadian rhythms, less blue light, sleep at dark and up to greet the sunrise;
- immune system support – eat phytochemicals, nutrient-dense food, touch soil, breathe in plants;
- physical health – move, breathe, feel calmer and collect vitamin D, inhale phytoncides.

HOW TO BUILD COMMUNITY THROUGH GARDENING

> *Each person, human or no, is bound to every other in a reciprocal relationship.*
> *Robin Wall Kimmerer*

We look after what we love, and we remember our communities in times of need. Perhaps because of recent catastrophic events, we are remembering a need for a kind of devotion to the Earth and to everything that is part of it. When we see nature as theophany – a revelation of our divine selves – it becomes more than a commodity to be exploited; we become part of the same community. Robin Wall Kimmerer's *Braiding Sweetgrass* explains how native Americans combine plant science with reciprocity, to build community with plants and animals. In the UK Sam Lee leads pilgrimages for the turtle dove and holds night-time concerts singing and playing in harmony with nightingales in woodland to bring back our connection to this more-than-human community. As growers, community brings us access to shared knowledge, plant exchanges and seed swaps. Friendships made over garden fences discussing dahlias and compost can become life-enhancing, even life-saving.

Community is a vital support to the farmers who grow our food, both to share knowledge and equipment and as a listening ear. In the UK the farming community network (FCN) offers confidential support to farming families – an important role in a community where young men suffer from three times the national average suicide rates. The gardening charity Perennial helps gardeners in hard times, and several other wonderful charities like Maggie's, Blackthorn Trust and Horatio's Garden nurture and heal people through contact with gardens.

Some things are more easily done as a community than alone. Cleaning our water, harvesting energy, growing food at scale and managing waste are a few. In the past, cities would have growers on the outskirts who would bring fresh vegetables in daily, and return carts would be laden with human waste to fertilise their fields. Most councils now collect green waste and food waste to be processed and sold as compost. The high temperatures used to ensure they are pathogen-free also kills the microbes, so, while useful as mulch, more life is needed

Opposite: As growers, community brings us access to shared knowledge, plant exchanges and seed swaps. Friendships made over garden fences discussing dahlias and compost can become life-enhancing, even life-saving.

The best way to regain reciprocity and resilience as a growing community is to use local seeds.

to create a good growing medium. Activist and passionate soil geek Michael Kennard set up the Compost Club in Lewes, a community subscription service that collects food waste from local homes in simple bokashi bins that he supplies. Michael processes this into living compost, which he partly returns to each home and partly sells to fund this community action.

Farmers' markets and box schemes are another great way to participate in community. By paying our farmers direct, we are building their autonomy and becoming part of the solution to the crisis in food health. Just knowing who grows our food and how it is grown is a big step towards having a closer relationship with the land and each other. The Riverford organic box scheme delivers 80,000 boxes a year in the UK to a community of regular customers. Founder Guy Singh-Watson has now sold to a co-owner trust of employees, who run and profit from it while in the community. The Wildfarmed movement in the UK is creating a community of regenerative growers, guaranteeing to buy mixed grains and pulses at a premium price, while supporting and providing training and workshops to their members.

It may not be practical to be fully food-sufficient in the UK but as growers we could be seed-sufficient. The best way to regain reciprocity and resilience as a growing community is to use local seeds; currently we import around 80 per cent of UK seed. Seeds from plants that are adapted to local conditions continue to evolve each season as climate develops.

Seeds work in close community with the soil. As the soil microbiome passes through the seed, knowledge is passed on to the next generation, which increases the chances of that seed surviving. Thriving plants produce more seed, so thriving seeds are vital to our long-term community resilience. In the UK the Gaia Foundation's Seed Sovereignty Programme has been working on a crowd breeding project with a group of eighteen farmers who collectively curate three crops. It is an experiment in adaptive farming – growing crops as diverse populations to allow genetics to roam more freely. By allowing new traits to occur and fully express themselves, the Gaia Foundation hopes to bring back diversity in four or five generations of plants. Gaia facilitator Holly Silvester is keen to stress the cultural and historical value of stable varieties and F1 hybrids in non-commoditised systems as well, but has found joyfulness in allowing what she calls promiscuous pollination, working in collaboration with nature, to produce disease- and weather-resilient crops.

Local seed hubs and seed-swap events are publicised on the internet as well as by plant fairs, horticultural groups and farmers' markets. As gardeners we can open our gardens for charity or organise compost collections, plant-division workshops and old-fashioned flower shows. There is a lot to be said for the like-minded communities we can find online, but nothing beats meeting on a street or in a village to share a plant chat and a cup of tea.

RECONNECTING WITH OURSELVES

> *The quieter you become, the more you are able to hear.*
> *Rumi*

Reconnecting with nature means reconnecting with ourselves. We can get wrapped up in busy minds, constantly thinking and planning. Try these ways to reconnect your mind to your body and land. List others you find helpful and keep it handy for times of stress. Take appropriate precautions to be safe, of course, like checking plants are edible and water is clean before drinking or swimming. See pages 230–234 for links and books.

Air

We can't survive long without air. Oxygen also clears our minds, while deep breaths slow our heart rate and bring us back to the moment. Take three large breaths as you wake. Releasing each one, recognise what you are grateful for and what you wish to leave behind. Imagine your breath being carried away by Mother Earth and recycled, like compost into roses. Try a walking meditation. Walk and breathe deeply. Notice everything around you.

Smell

Air is full of pheromones and signals. Ask what the scents you smell are signalling. How do they make you feel? Anxious, like burning tyres? Happy, like summer picnics or a grandmother's rose garden? Fill your garden and house with smells that make you feel good.

Wind

Listen to the wind. Wind makes distinct sounds depending on the trees it blows through. Winds vary, and can bring contrasting weathers. Try to tell the current wind or weather by sound before you open your eyes in the morning.

Water

Drink from springs where safe to do so. Wash your face in dew. Swim in lakes and rivers and the sea when clean and safe to do so. Take an outdoor shower in rainwater. Dance in the rain – naked if appropriate! Learn from this precious resource – water is resilient, shape shifting, purifying, life-giving. Add your own observations.

Practise yoga, qigong (shown here) or silent dance to bring mind and body together.

Challenge your concepts of time and value by looking at the exquisite details and life spans of insects.

Soil

Put your hands on the soil every day. If you are anxious, put them in the soil more often, breathe it in deeply. Notice the effect.

Plants

Note which volunteer plants are abundant in your garden. Are they a reflection of what you need, or of the soil condition, or both? Which minerals do they accumulate? What are they used for? Do you need that healing? Use a good reference book to forage leaves, flowers and roots to add to meals and to make teas for yourself and your plants.

Sunlight

Be outside for at least 20 minutes a day in daylight. Walk at sunrise. Give thanks at sunset. These are liminal spaces where magic happens.

Moonlight

Follow a moon calendar to understand the cycles, and how that links with your cycles and those around you. Walk in the moonlight for 20 minutes: avoid artificial light for 30 minutes beforehand – use a candle if you need to. If you are afraid of the dark – we are conditioned to be – go out before darkness and get used to it gradually.

Trees

Walk in the woods. Sit with your back to a tree. Choose a tree and notice it every day; what is it doing? How does it make you feel? Hug a tree (with permission) and ask to share its wisdom. Sometimes a thought or word will come to you; take note.

Flowers

Study a single flower. Smell it, feel it. If safe, have a nibble. Count parts like petals and sepals, male and female parts. Has it been pollinated? By what and how? Where does it grow naturally and in which plant archetypes? If there are plenty of this plant and it is not a protected species, dig one up and study its root structure. What are its medicinal properties and mythology?

Worms

Connect with worms! They are such inquisitive creatures. Channel your inner film-maker and watch a worm explore your hand – a *My Octopus Teacher* moment for you and the worm.

Insects

Take macro photos. Note the exquisite details. How long will this creature live in human time? Challenge your concepts of time and value.

Rescue a bee – put your hand in front of a bee stuck indoors and allow it to climb on to your fingers. Don't try to pick it up. Use a piece of paper if you are nervous or allergic. Take the bee outside to a flower and watch it drink nectar. Bees avoid stinging if possible as their internal organs get stuck in our skin and so they die. However, wasps can sting mammals more than once, so use a piece of paper and a jar for them. They are jumpy folk.

Watch a spider build a web. Read *Charlotte's Web.*

Which caterpillars like which plants? Watch a caterpillar engineer a leaf home. Which moth or butterfly will they metamorphose into?

Opposite: Use a good reference book to forage leaves, flowers and roots to add to meals and to make teas for yourself and your plants. Shown here: Camassia, Rosehips, hops, hawthorn, turkey tail and sloes.

Birds

Notice your birds – where do they perch and nest? What do they eat? Names aren't important, but they can help us to remember. Use an app to identify their songs.

Yourself

Use the wheel below to mark notable past events (photos on your phone make good prompts). Which events filled you with joy or anxiety, or other emotions? Which ones would you like to have more of? How can you make that happen?

Connect with your emotions: practice saying how you feel out loud without judgement. Say 'I feel cross' – or sad, happy or worried. Don't try to fix the emotion or event; just feel it and see how each emotion passes, like clouds across the sky.

Gaze at yourself in the mirror for a full minute. Describe your eyes to yourself. Send love to yourself. List all the things you like about yourself. Seal the list and put it where you'll find it later, or ask a friend to post it to you in six months' time.

Practise metta (loving kindness) meditation. Practise yoga, qigong or silent dance to bring mind and body together and ground them both. Walk with bare feet, feel the life below and all around. Breathe; kiss the Earth.

> *When we heal ourselves, we heal the world.*
> *Mark Nepo*

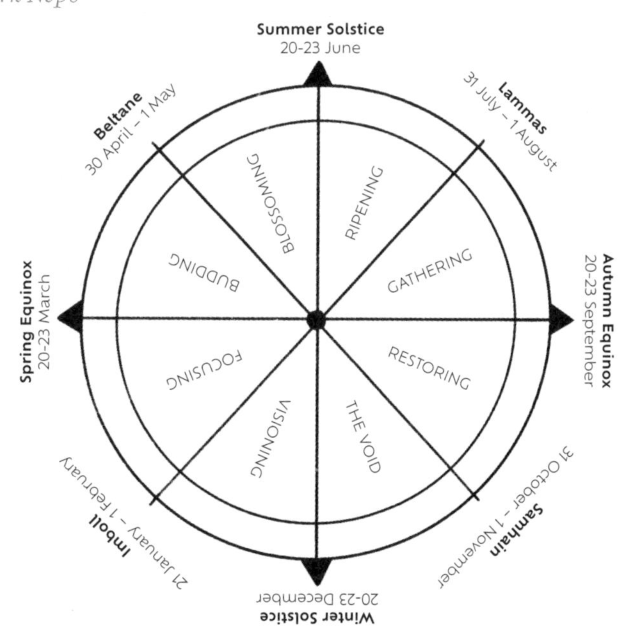

Right: Use this wheel of the main festivals of the year to mark notable past events. Which would you like to have more of? How can you make that happen?

SHARING KNOWLEDGE: A FINAL NOTE

> *Mystics understand the roots of the Tao but not its branches; scientists understand its branches but not its roots. Science does not need mysticism, and mysticism does not need science; but man needs both.*
> *Fritjof Capra*

If you take one thing from *The Kindest Garden*, I hope it is that we are all connected, from our soil biome to our gut biome to our cosmic biome. And that everything we can do to create health and happiness starts with kindness to ourselves. I hope you find good links on pages 230–4 to follow up the many questions raised, and that you will pursue those answers with endless enjoyment. The best advice I have received is to hold knowledge lightly, apply it without expectations of universal success, and work with each context.

I have referenced lots of science, and in these last chapters some quasi ('almost') science, since both are helpful, and cutting-edge thought is always going to be quasi-science until proven. Sensibly enough, the scientific establishment is wary of accepting things that we can see or feel but cannot prove. From discoveries like plate tectonics, the germ theory of disease and the existence of dark matter and dark energy, to scientists from Albert Einstein to Monica Gagliano and Elaine Ingham, all were dismissed at times when their truths did not fit with established knowledge.

A regenerative mindset is about being curious. It means working with your gut, your brain and your spirit, exploring anecdotal evidence as well as peer-reviewed evidence to create a better place for ourselves and our children's children. The regenerative agriculture movement has a saying, 'It's not the cow, it's the how', meaning our success depends on our approach. The way we manage our gardens, cities and landscapes is key to our healthy future. As forward-thinking *Homo sapiens* we are in pole position to use our own observations above all, together with our knowledge and mindsets, to make the most of nature's rich biology, minerals and energy.

Please take what serves you, and be curious about the rest. Assess where your land is now, establish your baselines, measure where you are with water and biodiversity, and see where you can get to in your own context. Notice your own emotions,

REGENERATIVE GARDEN CHECKLIST

Water

- ☐ Reduce mains water use and recycle grey water
- ☐ Capture rainwater: harvest from roofs to use on the land
- ☐ Slow runoff through planting and rainwater capture
- ☐ Capture water in ponds and swales
- ☐ Create an aquatic ecosystem
- ☐ Reduce garden water use, minimise irrigation needs through planting choices
- ☐ Use permeable hard surfaces, such as permeable paving or gravel

Soil

- ☐ Recycle organic matter on site, for example with compost bins, bokashi and wormery
- ☐ Use a no-dig approach
- ☐ Avoid compacting soil with heavy machinery
- ☐ Use an organic mulch to protect soil microbes and keep carbon and nutrients in the soil
- ☐ Do not use peat – create your own living compost
- ☐ Use green manures and compost teas

Plants and Trees

- ☐ Use layered planting to create a mini ecosystem
- ☐ Leave areas of longer grass habitat next to shorter
- ☐ Use plants to minimise building energy use, for example a vegetated roof or tree on the south side of a house for shade
- ☐ Use open-pollinated seeds, and save your own for genetic diversity
- ☐ Grow climbers on fences or lime mortar walls, plant native and edible hedges
- ☐ Grow vegetables, mix edible perennials with other planting
- ☐ Create climate resilient planting able to withstand drought and downpour
- ☐ Avoid planting non-native invasive plants and do not allow any to spread to countryside

Hard Landscape

- ☐ Source materials responsibly, ask questions about sources and check certifications like FSC timber
- ☐ Use reclaimed or reused materials, furniture and pots where possible
- ☐ Keep and reuse materials on site, for example crushed concrete for a dry garden
- ☐ Design for adaptability, for example ensure paving can be taken up for reuse
- ☐ Use locally sourced vernacular materials and craftspeople
- ☐ Maintain surfaces, furniture and structures to increase longevity but avoid microplastics and pollutants in paint, etc.

Ecology

- ☐ Include native plants
- ☐ Include a variety of forms of pollinator friendly species
- ☐ Include species which provide food for birds and small mammals, such as berry-producing shrubs
- ☐ Include shrubs and trees which provide shelter
- ☐ Install bird boxes or leave old dead wood trees
- ☐ Install bat boxes or leave crevices in roofs and walls
- ☐ Install insect habitats, such as insect posts, south-facing log piles and leave old wood *in situ*
- ☐ Install a hedgehog house, leave piles of brash and hedgehog holes in solid fences
- ☐ Install reptile hibernacula or leave large flat rocks for sunning spots
- ☐ Plan tasks to avoid disturbing wildlife at key times of the year (such as breeding season)
- ☐ Use low-level lighting with movement-detection or short timers to avoid confusing birds and bats
- ☐ Research local rare or endangered species and understand their needs
- ☐ Be a bit messy – allow dead plant matter to decay and scrub areas to develop as homes for wildlife

Pollution, Energy and Carbon Dioxide Emissions

- ☐ Use renewable energy for gardening needs, for example installing solar panels or raking instead of leaf blowing
- ☐ Cut down need for garden machinery in design stage, reducing energy use and pollution
- ☐ Source plants and materials locally

Pests, Diseases and Fertilisers

- ☐ Avoid chemical insecticides, pesticides and weedkillers, such as glyphosate
- ☐ Use organic approaches to pest and disease control, such as foliar sprays of compost tea for mildew and attracting predators for pests
- ☐ Use organic fertilisers, such as compost, comfrey or nettle tea instead of salt-based chemical fertilisers
- ☐ Use organic living compost, preferably made on site – extra points if tested for full soil food web

People

- ☐ Include areas for rest, reflection and celebration surrounded by plants and life
- ☐ Share skills, knowledge, seeds, plants, tools and fun with others

The best advice I have received is to hold knowledge lightly, apply it without expectations of universal success, and work with each context to make the most of nature's rich biology, minerals and energy.

energy and responses to food, to the land and to circumstances. Work with the seasons, moon cycles and circadian rhythms to increase energy flow for you and your land. If you find these ideas useful, please share them with friends, question them, experiment and make your own evidence for your own context. Above (page 225) is a list as an aide-memoire (not a 'to do' list!) for some of the ideas I have covered.

Writing this at the end of summer, with the swallows preparing to leave, brings home my pleasure in the long days and quiet dawns of the northern hemisphere. I have been gathering vitamin D through the sun on my skin along with vitamin C in stores of rosehips, sea buckthorn (*Hippophae rhamnoides*) and elder (*Sambucus nigra*) berries plus medicinal mushrooms and birch polypore honey to see me through the shorter days of winter. This connection to the land by working with the seasons is useful physically and mentally too. I am preparing for the next phase with less sun and fewer flowers and insects, but more time to make, to bake and mend, to study and work. And then spring will come again and with it a surge of energy from the sun to the plants and to the soil, to the animals and to you and me. A miracle every time.

> *What wisdom can you find that is greater than kindness?*
> *Jean-Jacques Rousseau*

APPENDIX 1
COMPARISON OF POND LINERS

	SYNTHETIC LINERS			CONCRETE	CLAY		NO LINER
	Butyl rubber liner	PVC plastic liner	HDPE plastic liner	Concrete liner	Puddled clay	Bentomat (geosynthetic clay liner)	Excavation into impermeable geology
Wildlife value							
Global warming (carbon impact)							
Installation difficulty							
Ease of shaping							
Maintenance level							
Longevity							
Strength							
Watertightness							
Temperature stability							
UV resistance							
Ozone resistance							
Chemical resistance (oils and solvents)							
Cost							
Karma rating							

APPENDIX 2
GREY WATER RECYCLING

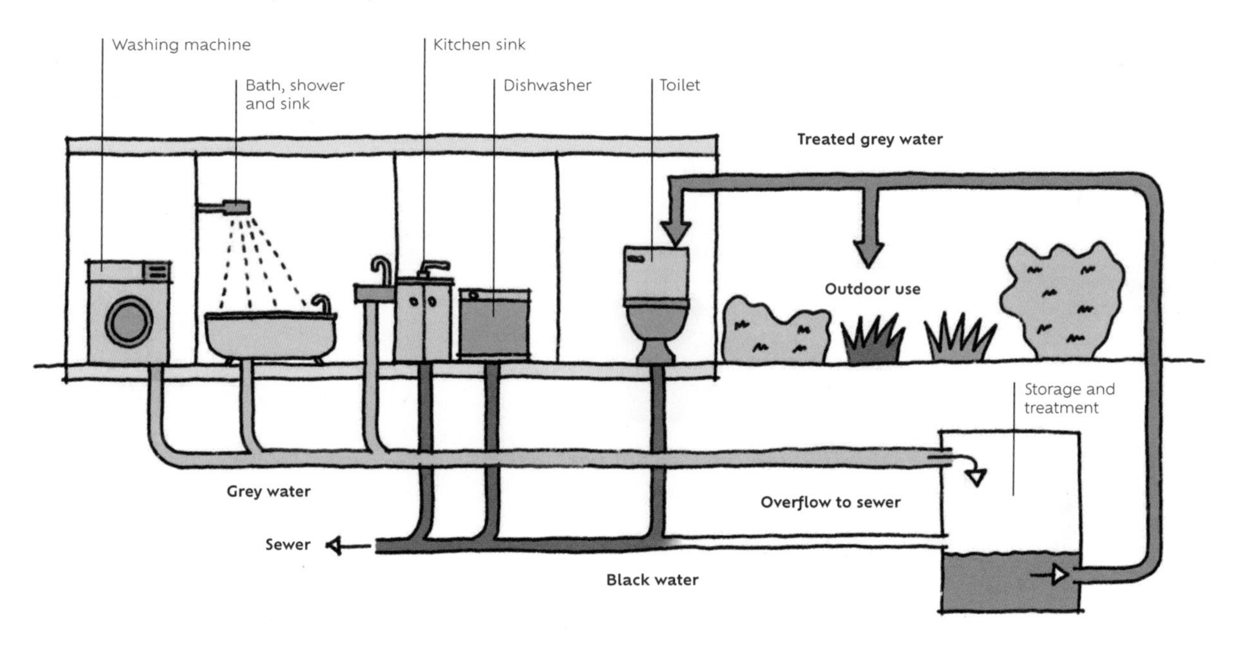

RESOURCES AND FURTHER INFORMATION

WILD FLOWER SEEDS

British Wild Flower Seeds: https://www.wildflowers.uk
Emorsgate Seeds: https://wildseed.co.uk
Heritage Seeds: http://www.heritageseeds.co.uk
John Chambers Wildflowers: https://www.johnchamberswildflowers.co.uk
Kent Wildflower Seeds: https://kentwildflowerseeds.co.uk
Wildflower Turf: https://wildflowerturf.co.uk

PERENNIAL VEGETABLES

Edulis: https://www.edulis.co.uk/collections/perennial-vegetables
Incredible Vegetables: https://incrediblevegetables.co.uk
Otter Farm: https://www.otterfarm.co.uk product-category/vegetables/perennial-vegetables

OPEN-POLLINATED SEEDS

Baddaford Farm Collective: https://www.baddaford.co.uk/collective/vital-seeds
Seed Sovreignty: https://www.seedsovereignty.info/resources/where-to-buy-open-pollinated-agro-ecological-seed

NURSERIES

Edible Culture: https://edibleculture.co.uk
Hillier Trees: https://trees.hillier.co.uk
How Green Nursery Ltd: https://howgreennursery.co.uk
Incredible Vegetables: https://incrediblevegetables.co.uk
Keepers Nursery: https://www.keepers-nursery.co.uk
Marchants Gardens & Nursery: https://www.marchantshardyplants.co.uk
New Wood Trees: https://newwoodtrees.co.uk
Organic Bulbs: https://www.organicbulbs.com
Palmstead: https://www.palmstead.co.uk
Sunnyside Rural Trust: https://www.sunnysideruraltrust.org.uk

VEGETABLE BOXES

Abel & Cole: https://www.abelandcole.co.uk
East Coast Organics: https://www.eastcoastorganics.co.uk
Farm Direct: https://farm-direct.com
Keveral Farm: http://www.keveral.org
The Organic Pantry: https://www.theorganicpantry.co.uk
Riverford: https://www.riverford.co.uk
The Veg Box Company: https://thevegboxcompany.co.uk

RIVERS AND RIVER RESTORATION

Earth Law Center https://www.earthlawcenter.org
Environment Agency Owning a Watercourse Guidance https://www.gov.uk/guidance/owning-a-watercourse
Environment Agency Statutory Main River Map: https://arcg.is/10rSDT0
The National River Flow Archive: https://nrfa.ceh.ac.uk/about-nrfa
The Rivers Trust: https://theriverstrust.org/
Surfers Against Sewage: https://www.sas.org.uk/
Pollock, M.M. et al., 'River Restoration and Meanders' in *Ecology and Society* 19(4), 43 (2014)
Puttock, A. et al., 'Making Space for Water: Investing in Nature-based Solutions with Beavers', in *EGU General Assembly*, Vienna, Austria (14–19 April, 2024), https://doi.org/10.5194/egusphere-egu24-4650

ECOSYSTEM ENGINEERS

Brazier, Richard E., et al., 'Beaver: Nature's ecosystem engineers', in *Wiley Interdisciplinary Reviews: Water* 8.1 (2021)

Davidson, D.W. et al., 'Mutualistic ants as an indirect defence against leaf pathogens', in *Nature Plants* 3(1), 1–7 (2017)

Edwards, Clive A. and Bohlen, P.J., *Biology and Ecology of Earthworms* (Springer Netherlands, 1996)

Jones, C.G., Lawton, J.H. and Shachak, M., 'Organisms as ecosystem engineers', in *Oikos* 69, 373–386 (1994)

Jones, C.G. and Lawler, S.P., 'Beavers as landscape engineers: the manipulation of water bodies by a keystone species', in *Reviews in Environmental Science and Biotechnology* 14(4), 333–347 (2015)

A SELECTION OF KEY BOOKS

Adams, Douglas, *The Hitchhiker's Guide to the Galaxy* (Pan Books, 1979)

Albrecht, William A., *Albrecht's Foundation Concepts* (Acres USA Inc., 1975)

Andersen, Arden B., *Science in Agriculture: Advanced Methods for Sustainable Farming* (Acres USA Inc., 2000)

Armstrong, Karen, *Sacred Nature: How We Can Recover Our Bond with the Natural World* (Random House, 2022)

Averis, Ben, *Plants and Habitats: An Introduction to Common Plants and Their Habitats in Britain and Ireland* (Ben Averis, 2013)

Bartholomew, Alick, *Hidden Nature: The Startling Insights of Viktor Schauberger* (Floris Books, 2014)

Benton, Ted and Owens, Nick, *Solitary Bees* (Collins New Naturalist Library, 2023)

Berry, Wendell, *The Peace of Wild Things* (Penguin Books, 2018)

Brown, Gabe, *Dirt to Soil: One Family's Journey into Regenerative Agriculture* (Chelsea Green Publishing, 2018)

Bruton-Seal, Julie and Seal, Matthew, *Hedgerow Medicine: Harvest and Make Your Own Herbal Remedies* (Merlin Unwin Books, 2008)

Bruton-Seal, Julie and Seal, Matthew, *Wayside Medicine: Forgotten Plants and How to Use Them* (Merlin Unwin Books, 2017)

Callahan, Philip S., *Tuning in to Nature: Solar Energy, Infrared Radiation, and the Insect Communication System* (Devin-Adair Company, 1975)

Calvo, Paco and Lawrence, Natalie, *Planta Sapiens: Unmasking Plant Intelligence* (Little, Brown, 2022)

Capra, Fritjof, *The Web of Life: A New Scientific Understanding of Living Systems* (Anchor Books, 2016)

Conway, Ed, *Material World: A Substantial Story of Our Past and Future* (Ebury Publishing, 2023)

Crawford, Martin, *Creating a Forest Garden: Working with Nature to Grow Edible Crops* (Bloomsbury, 2010)

Eisenstein, Charles, *The More Beautiful World Our Hearts Know is Possible* (North Atlantic Books, 2013)

Enders, Giulia, *Gut: The Inside Story of Our Body's Most Underrated Organ* (Greystone Books, 2015)

Falk, Steven J., *Field Guide to the Bees of Great Britain and Ireland* (Bloomsbury, 2015)

Fiennes, Jake, *Land Healer: How Farming Can Save Britain's Countryside* (Ebury Publishing, 2022)

Filippi, Oliver, *The Dry Gardening Handbook: Plants and Practices for a Changing Climate* (Thames & Hudson, 2008)

Fukuoka, Masanobu, *The Natural Way of Farming: The Theory and Practice of Green Philosophy* (Japan Publications, 2009)

Fukuoka, Masanobu, *The Dragonfly Will Be the Messiah* (Penguin Books, 2021)

Gagliano, Monica, Thus Spoke the Plant: A Remarkable Journey of Groundbreaking Scientific Discoveries and Personal Encounters with Plants (North Atlantic Books, 2018)

Garnier, Eric, Navas, *Marie-Laure and Grigulis, Karl, Plant Functional Diversity: Organism Traits, Community Structure, and Ecosystem Properties* (Oxford University Press, 2015)

Goldsmith, Ben, *God Is An Octopus: Loss, Love and a Calling to Nature* (Bloomsbury USA, 2024)

Gow, Derek, *Bringing Back the Beaver: The Story of One Man's Quest to Rewild Britain's Waterways* (Chelsea Green Publishing, 2020)

Harman, Hattie and Williams, Joe Jack, *Materials: An Environmental Primer* (RIBA Publishing, 2024)

Hawken, Paul, *Regeneration: Ending the Climate Crisis in One Generation* (Penguin Books, 2021)

Howard, Albert, *The Soil and Health: Farming & Gardening for Health or Disease* (Rodale Books, 1947)

Imms, Augustus Daniel, *Insect Natural History* (Collins New Naturalist Library, 2017)

Keble Martin, William, *The Concise British Flora in Colour* (Ebury Press, 1969)

Kempf, John, *Quality Agriculture: Conversations about Regenerative Agronomy with Innovative Scientists and Growers* (Regenerative Agriculture Publishing, 2020)

Kumar, Satish, *Soil, Soul, Society: A New Trinity for Our Time* (Parallax Press, 2024)

Lake, Sophie, Liley, Durwyn, Still, Robert and Swash, Andy, *Britain's Habitats: A Field Guide to the Wildlife Habitats of Great Britain and Ireland* (Fully Revised and Updated Second Edition, Princeton University Press, 2015)

Langford, Sarah, *Rooted: Stories of Life, Land and a Farming Revolution* (Penguin Books, 2022)

Lovelock, James, Novacene: The Coming Age of Hyperintelligence (Penguin Books, 2019)

Macdonald, Benedict, *Rebirding: Rewilding Britain and Its Birds* (Pelagic Publishing, 2019)

Massy, Charles, *Call of the Reed Warbler: A New Agriculture, a New Earth* (Charles Green Publishing, 2018)

Masters, Nicole, *For the Love of Soil: Strategies to Regenerate Our Food Production Systems* (Integrity Soils Ltd, 2019)

McCaman, Jay L., *When Weeds Talk* (Acres USA Inc., 2013)

Monbiot, George, *Regenesis: How to Feed the World Without Devouring the Planet* (Penguin Books, 2022)

Montgomery, David R. and Biklé, Anne, *What Your Food Ate* (W.W. Norton, 2022)

Nepo, Mark, *The Exquisite Risk: Daring to Live an Authentic Life* (Harmony Books, 2005)

Newman Hugh, Rocka, Jewels and Creightmore, Richard, *Geomancy: Earth Grids, Ley Lines, Feng Shui, Divination, Dowsing, and Dragons* (Wooden Books, 2021)

Nhất Hạnh (Thích.), *The Miracle of Mindfulness: A Manual on Meditation* (Beacon Press, 1987)

O'Connell, Marina, *Designing Regenerative Food Systems: And Why We Need Them Now* (Hawthorn Press, 2022)

O'Donohue, John and Quinn, John, *Walking in Wonder: Eternal Wisdom for a Modern World* (Random House 2018)

Palmer, Nigel, *The Regenerative Grower's Guide to Garden Amendments: Using Locally Sourced Materials to Make Mineral and Biological Extracts and Ferments* (Chelsea Green Publishing, 2020)

Pollock, Gerald H., *The Fourth Phase of Water: Beyond Solid, Liquid, and Vapor* (Ebner & Sons, 2013)

Polunin, Nicholas, *Introduction to Plant Geography and Some Related Sciences* (Longmans, 1964)

Polunin, Oleg and Walters, Martin, *A Guide to the Vegetation of Britain and Europe* (Oxford University Press, 1985)

Polunin, Oleg, *Flowers of Europe: A Field Guide* (Oxford University Press, 1969)

Proctor, Michael, Yeo, Peter and Lack, Andrew, *The Natural History of Pollination* (Collins New Naturalist Library, 2009)

Rackham, Oliver, *The Illustrated History of the Countryside* (Weidenfeld & Nicolson, 2008)

Rackham, Oliver, *Trees and Woodland in the British Countryside* (Orion, 2020)

Raskin, Ben, *The Woodchip Handbook: A Complete Guide for Farmers, Gardeners and Landscapers* (Chelsea Green Publishing, 2021)

Rew, Kate, *The Outdoor Swimmers' Handbook* (Ebury Publishing, 2022)

Rodwell, John S., *British Plant Communities: Volume 3, Grasslands and Montane Communities* (Cambridge University Press, 1991)

Rose, Francis and O'Reilly, Clare, *The Wild Flower: Key How to Identify Wild Flowers, Trees and Shrubs in Britain and Ireland* (Revised Edition, Frederick Warne, 2006)

Rumi, *The Essential Rumi*, translated by Coleman Barks (HarperCollins, 1995)

Savory, Allan and Butterworth, Jody, *Holistic Management: A Commonsense Revolution to Restore Our Environment* (3rd edition, Island Press, 2016)

Schofield, Lee, *Wild Fell: Fighting for Nature on a Lake District Hill Farm* (Transworld, 2022)

Sheldrake, Merlin, *Entangled Life: How Fungi Make Our Worlds, Change Our Minds & Shape Our Futures* (Random House, 2020)

Sheldrake, Rupert, *Morphic Resonance: The Nature of Formative Causation* (Inner Traditions/Bear, 2009)

Shrubsole, Guy, *The Lost Rainforests of Britain* (HarperCollins, 2022)

Sjöman, Henrik and Anderson, Arit, *The Essential Tree Selection Guide: For Climate Resilience, Carbon Storage, Species Diversity and Other Ecosystem Benefits* (Filbert Press, 2013)

Spyri, Johanna, *Heidi* (Illustrated Edition, HarperCollins, 1996)

Stamets, Paul, *Fantastic Fungi: How Mushrooms Can Heal, Shift Consciousness, and Save the Planet* (Earth Aware Editions, 2019)

Stika, Jon, *A Soil Owner's Manual: How to Restore and Maintain Soil Health* (CreateSpace Independent Publishing Platform, 2016)

Tree, Isabella, *Wilding: The Return of Nature to a British Farm* (Pan Macmillan, 2018)

Wohlleben, Peter, *The Power of Trees: How Ancient Forests Can Save Us If We Let Them* (Greystone Books, 2023)

Wulf, Andrea, *The Invention of Nature: The Adventures of Alexander Von Humboldt, the Lost Hero of Science* (John Murray Press, 2015)

Wyndham, John, *The Day of the Triffids* (Michael Joseph, 1951)

Young, Rosamund, *The Secret Life of Cows* (Faber & Faber, 2017)

PODCASTS

Advancing Eco Agriculture (John Kempf)
Emergence Magazine podcast
Hope Springs (Annabel Heseltine for Resurgence magazine)
Investing in Regenerative Agriculture and Food (Koen van Seijen)
ReGenAgChat (Liz Genever and Nic Renison)
Rewild the World (Ben Goldsmith)
Tara Brach

ONLINE RESOURCES AND INFORMATION

Marian Boswall Landscape Architects
Detailed information and data on a wide range of topics, including soil, water, materials and trees can be found in the 'Insight' section of our website. These resources are updated regularly: https://www.marianboswall.com/insight

Acres USA
Publisher of books on sustainable and organic farming
https://bookstore.acresusa.com

Agroforestry Research Trust
Courses and education
https://www.agroforestry.co.uk

BRE Group
Environmental Product Declarations (EPDs)
https://bregroup.com/services/testing-certification-verification/en-15804-environmental-product-declarations

Bamboo Sticks Exercises
June Mitchell demonstrates qigong exercises
https://www.youtube.com/watch?v=-pluOjzBHS4

British Geological Survey
Research and information
https://www.bgs.ac.uk

Carbon Calling
Regenerative farming conference
https://www.carboncalling.farm

Compost Club
Soil courses and workshops
https://www.compostclub.online

The International EPD® System
Environmental Product Declarations (EPDs)
https://www.environdec.com/home

Integrity Soils (Nicole Masters)
Agroecology coaching and training
https://integritysoils.com

The Land Gardeners
Soil education
https://www.thelandgardeners.com

Life Sustaining Way of the Heart
Qi Gong Bamboo Stick Exercises Booklet
https://plumvillage.shop/products/gifts/journals-cards-calendars-prints/bamboo-stick-exercises-booklet

National Library of Scotland
Historic maps of the UK
https://maps.nls.uk/

NBN Atlas
UK biodiversity data
https://nbnatlas.org

NHBS (Natural History Book Service)
Mail-order ecology and wildlife books
https://www.nhbs.com

Parliament of Religions
Confucian statement on ecology
https://parliamentofreligions.org/articles/confucian-statement-on-ecology

Phys.org
The evolution of oak trees in the Anthropocene
https://phys.org/news/2022-01-evolution-climate-oaks-rapidly-anthropocene.html

Sprout Club
Sustainable growing community
https://www.sproutclub.com

Soil & Health Library
Free downloadable eBooks
https://soilandhealth.org

Soil Ecology Lab
Soil biology testing and training
https://soilecologylab.co.uk

Soil Food Web (Dr Elaine Ingham)
Courses in soil health
https://www.soilfoodweb.com

Soilmentor
Soil monitoring app
https://soils.vidacycle.com

Soil Redemption (Perry Hanbury)
Soil-lab assessments
https://www.soilredemption.com

Wageningen University & Resarch
Archive of Root System Drawings
https://images.wur.nl/digital/collection/coll13

Wildfarmed
Flour produced using regenerative farming
https://wildfarmed.com

REGENERATIVE RETREATS AND LEARNING CENTRES

42 Acres
Regenerative retreat in Somerset
https://www.42acres.com

Apricot Centre
Sustainable farm and wellness centre in Devon
https://www.apricotcentre.co.uk

Avalon Wellbeing
Wellbeing retreat centre at Broughton Hall Estate in Yorkshire
https://www.avalonwellbeing.com

Badgells Wood
Off-grid woodland camping in Kent
https://www.badgellswoodcamping.co.uk

Elmley Nature Reserve
Nature reserve with accommodation in Kent
https://www.elmleynaturereserve.co.uk

Groundswell
Regenerative Agriculture Festival in Hertfordshire
https://groundswellag.com/

Rewilding Coombeshead (Derek Gow)
Nature breaks in Devon
https://rewildingcoombeshead.co.uk/stay-with-us

Rothamstead Research
Laboratories and experimental farms across four locations in England
https://www.rothamsted.ac.uk

Schumacher College
Ecology college in Devon
https://campus.dartington.org/schumacher-college-2024/

INDEX

ACKNOWLEDGEMENTS

You cannot just be by yourself alone. You have to inter-be with every other thing. This sheet of paper is, because everything else is.
Thích Nhất Hạnh

This book is a synthesis of so many things I have learnt and discovered from other great people that I am only a small part shareholder in it at all. I have also had enormous help from several wonderful people, and small nuggets of vital insight from others. I have very tolerant friends who put up with me disappearing into rabbit holes for months, and a fantastic team in the studio. Here are just a few thank yous, along with my gratitude to so many:

Rupert Boswall
Marc Boughton and Steph Hill
Jim and Sheila Briggs
Andy Cato
Joanna Chisholm
Donna Cox
Edward Davenport
Toby Diggens
David Dodd and the Outdoor Room team
Phyllis Estlick
Tecwyn Evans
Sasha Georgiev
Ben Goldsmith
Derek Gow
Alice Graham, Michael Brunström, Glenn Howard, Liz Somers and the fantastic team at Frances Lincoln
Elaine Ingham and the Soil Food Web team
Jason Ingram, incredible photographer, friend and guide
Caroline Jackson
John Kempf
Jean Lawson
Nicole Masters and the CREATE team
June Mitchell and Satish Kumar
Charlotte Molesworth
Sheena and Paul Murphy
Angela Nevill
Guy Nevill
Sophie Nevill for the title inspiration
Marina O'Connell
Nick Padwick
Eliza Pelham and Ed Conway
Andy Phillips
Sophie Pollock and Hamish Cleland for lovely illustrations
Kim Polman
Alan Puttock
Robyn and Mark Reeves
Archie Ruggles-Brise
Holly Silvester
Seth Tabatznik and Renata Minerbo, Russell Rigler and the 42 Acres team
Naomi Warman
Guy, Oliver, Rachel and Will Watson at Riverford
Joe White for diagrams and extra research

and the many authors of books, and articles, podcasts and meditations, some of which are listed on pages 230–4.

Quarto

First published in 2025 by Frances Lincoln,
an imprint of The Quarto Group.
One Triptych Place, London, SE1 9SH,
United Kingdom
T (0)20 7700 9000
www.Quarto.com

EEA Representation, WTS Tax d.o.o., Žanova ulica 3,
4000 Kranj, Slovenia

Text copyright © 2025 Marian Boswall
Plans and diagrams © 2025 Marian Boswall
Photographs © 2025 Jason Ingram except for
Marian Boswall: pages 15, 38, 71, 81, 112, 152, 155, 156, 160, 194, 195, 206, 208, 220; Alan Puttock: pages 20, 74; Shutterstock: pages 106 (Erik Mandre), 109 above (zakharov aleksey), 109 below (Homestudio 2), 189 (Budimir Jevtic)
Design copyright © 2025 Quarto Publishing plc

Marian Boswall has asserted her moral right to be identified as the Author of this Work in accordance with the Copyright Designs and Patents Act 1988.

All rights reserved. No part of this book may be reproduced or utilised in any form or by any means, electronic or mechanical, including photocopying, recording or by any information storage and retrieval system, without permission in writing from Frances Lincoln.

Every effort has been made to trace the copyright holders of material quoted in this book. If application is made in writing to the publisher, any omissions will be included in future editions. A catalogue record for this book is available from the British Library.

ISBN 978-0-7112-8943-7
Ebook ISBN 978-0-7112-8944-4

10 9 8 7 6 5 4 3 2 1

Design by Glenn Howard

Publisher: Philip Cooper
Commissioning Editor: Alice Graham
Senior Editor: Michael Brunström
Editorial Assistant: Izzy Toner
Senior Designer: Isabel Eeles
Production Controller: Alex Merrett

Printed in China